A Practical Manual on

Microbiologically Influenced Corrosion

Gregory Kobrin,

Editor

Published by

NACE International
1440 South Creek Drive
Houston, TX 77084

Carolyn Donahoo
Director, Publications Division

Aunee Pierce
Senior Editor, Books

Cicely Wynne
Editor, Books

Kenneth Martinez
Production Artist

ISBN 1-877914-56-8

Library of Congress Catalog Card Number: 93-85751

T-3J-7 Members, Manual Contributors, and Reviewers

The scope of Unit Committee T-3J, Microbiologically Influenced Corrosion, is to provide a cooperative mechanism for studies in the area of corrosion caused by microbiological organisms. Specific activities will include:

- Producing reports on state of the art,
- Identifying and publishing case histories,
- Sponsoring symposia,
- Developing analytical techniques,
- Providing practical mechanisms,
- Determining basic mechanisms, and
- Defining needed research.

Current chairman is Gregory Kobrin. Immediate past chairman is Arthur J. Freedman.

Task Group T-3J-7 was formed during the CORROSION/83 Conference in Anaheim, California, in April 1983 to write a textbook on microbiologically influenced corrosion (MIC) over a five-year period. In a meeting of the Task Group during CORROSION/85 in Boston, Massachusetts, in March 1985, the group decided that a definitive text on MIC was premature, but that a practical manual might be appropriate. Kobrin agreed to chair the group with the revised charter in mind. However, the manual concept and format, with author assignments and so forth, were not officially developed until the T-3J-7 meeting during Fall Committee Week/86 in Atlanta, Georgia, in September 1986. Only minor modifications to the format and contents have been made since.

The manual is designed to provide practical information on: the roles microorganisms such as bacteria and fungi play in corrosion reactions; the occurrence of MIC in several industries; how to go about identifying whether or not a particular corrosion problem has been influenced by microorganisms; MIC treatment and prevention tips; selected case histories from the technical literature with photographs in color plus an additional "catalog" of color photos to further assist the reader in diagnosing potential MIC; and selected references for further study. Authors of the various sections are known internationally as experts in this field.

We believe this approach to the subject of MIC is unique and that this manual will be a valuable addition to the technical literature. At the time of its publication, T-3J-7 already will have initiated work on a second edition that will include an expanded section on occurrence of MIC in industry and an update of other sections.

T-3J-7 Members

Peter William Allison
Susan W. Borenstein
Thomas C. Breske
Robert Brown
John R. Bruce
Paul A. Burda
James R. Burge
Patricia A. Burke
Allan P. Castillo
Bruce A. Cookingham
Michael Davies
Stephen C. Dexter
Arthur J. Freedman
Edward Freiter
Paige A. Herbert
Gary Horacek
Doug Jones
Jeff Kearns
Ernest W. Klechka
Gregory Kobrin
John J. Kvochak
G.J. Licina
Emanuel S. Littmann[1]
Richard W. Lutey
Rupi Prasad
Alan L. Smith
Arthur A. Stein
John G. Stoecker
Robert E. Tatnall
Hector A. Videla
Christopher L. Wiatr
John F. Wilkes
Donald P. Wyman

T-3J-7 Contributors

Susan W. Borenstein
Stephen C. Dexter
R.G.Y. Edyvean
Gill G. Geesey
Gary Horacek
Gregory Kobrin
Emanuel S. Littmann[1]
Richard W. Lutey
Y. Mussalli
D.H. Pope
H.W. Rossmoore
L.A. Rossmoore
Arthur A. Stein
John G. Stoecker
Robert E. Tatnall

T-3J-7 Reviewers

Arthur F. Brunn
Stephen C. Dexter
John Fraser
Gregory Kobrin
P.B. Lindsay
Arthur A. Stein
John G. Stoecker
Christopher L. Wiatr

[1]Deceased.

Aerobe: An organism that requires oxygen to live.

Aerobic: Exposed to oxygen.

Algae: A heterogeneous group of photosynthetic plants, ranging from microscopic single-cell forms to multicellular very large forms such as seaweed.

Anaerobe: An organism that lives without oxygen.

Anode: The electrode of an electrolyte cell at which oxidation occurs. Electrons flow away from the anode in the external circuit. Corrosion occurs at this electrode, and metal ions enter the solution.

Bacteria (Bacterium): Any of a large group of microscopic organisms that are often aggregated into colonies, are enclosed by a cell wall or membrane (procaryotes), and lack fully differentiated nuclei. Bacteria may exist as free-living organisms in soil, water, or organic matter, or as parasites in the live bodies of plants and animals.

Biocide: A compound toxic to any living thing.

Biomass: Organic matter grown by the photosynthetic conversion of solar energy.

Cathode: The electrode of an electrolytic cell at which reduction takes place. Electrons flow toward the cathode in the external circuit. Typical cathodic processes are cations taking up electrons and being discharged, oxygen being reduced, and the reduction of a metallic element or group of elements from a higher to a lower valence state.

Cathodic Polarization: The change of the electrode potential in the active (negative) direction caused by current flow (see Polarization).

Cathodic Protection: Reduction of corrosion rate by shifting the potential of the corroding metal toward a less oxidizing potential with application of an external electromotive force, such as with a less noble metal (galvanic) anode or an impressed direct current.

Colony: A visible growth of microorganisms on a medium.

Concentration Cell: A corrosion cell in which the driving force is caused by a difference in concentration of a component in the electrolyte.

Contaminant: Any unwanted or undesired material in an environment.

Corrosion Potential (E_{corr}): The potential of a corroding surface in an electrolyte relative to a reference electrode. Also called rest potential, open circuit potential, freely corroding potential.

Corrosion Rate: The rate at which corrosion proceeds, expressed as either weight loss or penetration per unit time.

Corrosion Resistance: Ability of a metal to withstand corrosion in an environment.

Corrosion: The deterioration of a material, usually a metal, by reaction with its environment.

Corrosiveness, Corrosivity: The tendency of an environment to cause corrosion in a given corrosion system.

Crevice Corrosion: A form of localized corrosion occurring at locations where easy access to the bulk environment is prevented, such as at the mating surfaces of metals or assemblies of metals and nonmetals (see Concentration Cell).

Critical Pitting Potential (E_{cp}, E_p, E_{pp}): The lowest value of oxidizing potential at which pits nucleate and grow.

Culture Medium: Any nutrient system for the artificial cultivation of bacteria or other microorganisms, usually a mixture of organic and inorganic materials.

Depolarization: The removal of factors resisting the flow of current in an electrochemical cell.

Deposit Attack: Corrosion occurring under or around a deposit on a metallic surface (also called poultice corrosion).

***Desulfovibrio*:** A genus of bacteria that reduces sulfate to H_2S and obtains energy by oxidation of H_2S or organic molecules.

Differential Aeration Cell: A corrosion cell caused by differences in oxygen concentration in an electrolyte.

Electrochemical Cell: An electrochemical system consisting of an anode and a cathode in electrical contact and immersed in an electrolyte. The anode and cathode may be different metals or dissimilar areas on the same metal surface.

Electrolyte: A chemical substance or mixture, usually liquid, containing ions that migrate in an electric field.

Environmental Cracking: Fracture of normally ductile metal in which the corrosive effect of the environment is a causative factor. Environmental cracking is a general term that includes all of the following terms:

- Corrosion Fatigue
- High Temperature Hydrogen Attack
- Hydrogen Blistering
- Hydrogen Embrittlement
- Hydrogen-Induced Cracking
- Liquid Metal Cracking
- Stress Corrosion Cracking
- Sulfide Stress Cracking

Enzyme: Organic catalyst of biochemical reactions in a bacterial cell; composed of protein.

Erosion: Destruction of materials by the abrasive action of moving fluids, usually accelerated by the presence of solid particles.

Exopolymer: An extracellular material, frequently sticky, that defines the shape of organisms and serves to trap particulate matter.

Facultative Cells: Bacterial cells that can live either in the presence or absence of oxygen.

Ferrobacteria: Any of a group of bacteria that oxidize iron as a source of energy. The oxidized iron, in the form of $Fe(OH)_3$, is then deposited in the environment by secretion from the bacterium. The energy obtained from these reactions is used to carry on processes in which the basic substances needed by the bacterium are manufactured. These bacteria, also known as iron bacteria, are commonly found, for example, in seepage waters of coal and iron mining areas where iron compounds abound.

Fouling: An accumulation of deposits on a surface. This term includes accumulation and growth of marine organisms on a submerged metal surface and also includes the accumulation of deposits (usually inorganic) on heat exchanger tubing.

Fungus (Fungi): Any of a major group of parasitic plants that lack chlorophyll, including molds, mildews, smuts, mushrooms, and yeasts.

Galvanic Anode: A metal that, because of its relative position in the galvanic series, provides sacrificial protection to metals that are more noble in the series when coupled in an electrolyte.

Galvanic Corrosion: Accelerated corrosion associated with current resulting from electrical coupling of dissimilar metals in an electrolyte.

General Corrosion: A form of deterioration that is distributed more or less uniformly over a surface.

Habitat: The natural environment of an organism.

Iron Bacteria: See Ferrobacteria.

Medium: A substance used to provide nutrients for cell growth. It may be liquid (broth) or solid (agar).

Metabolism: The entire set of enzyme-catalyzed transformations of organic nutrient molecules in living cells.

Metabolite: A chemical intermediate in the enzyme-catalyzed reactions of metabolism.

Microbe: A microscopic organism; applied particularly to bacteria.

Microbiocide: Any chemical that will kill microorganisms. Used synonymously with biocide and bacteriocide.

Microbiology: The science dealing with the structure, classification, physiology, and distribution of microorganisms, and with their technical and medical significance. The term microorganism is applied to the unicellular and structurally closely related simple representatives of the plant and animal kingdoms. With few exceptions, unicellular organisms are invisible to the naked eye and generally have dimensions between a fraction of a micron and 200 microns.

Microorganism: Any organism of microscopic size that requires a microscope to be seen.

Obligate: A term used to describe a metabolism restricted to a similar type, as in the expression, obligate anaerobe.

Passivation: A reduction of the corrosion rate of a metal due to formation of a protective film, usually an oxide, on its surface.

Pits, Pitting: Localized corrosion of a metal surface that is confined to a small area and takes the form of cavities.

Plankton (Planktonic): Microorganisms floating or drifting in a body of water.

Polarization: The deviation from the open circuit potential of an electrode resulting from the passage of current.

Rust: Corrosion product consisting primarily of hydrated iron oxide; a term properly applied only to iron and ferrous alloys.

Scaling: The deposition of water-insoluble constituents on a metal surface.

Sessile: Attached to a surface.

Slime: An extracellular material produced by some microorganisms, characterized by a slimy consistency of varied chemical composition.

Sterile: Free of any living organisms.

Strain: Microorganisms of the same species but with distinctive characteristics and not usually considered a different breed or variety.

Sulfate-Reducing Bacterium (SRB): Any organism that metabolically reduces sulfate to H_2S; includes a variety of microorganisms (see *Desulfovibrio*).

Thermophile (Thermophilic): An organism that grows best at temperatures greater than 50°C (122°F).

Tuberculation: The formation of localized corrosion products scattered over the surface in the form of knoblike mounds called tubercles.

Underfilm Corrosion: Corrosion that occurs under organic films in the form of randomly distributed threadlike filaments or spots.

Yeast: A fungus of the family *Saccharomycetaceae* used especially in the making of alcoholic liquors and as leavening in baking.

Bibliography

1. *NACE Corrosion Engineer's Reference Book* (Houston, TX: NACE, 1991).
2. "The Language of Biotechnology—A Glossary" (Phillips Petroleum Co. Research Center, 1984).

Introduction

By R.E. Tatnall

Corrosion Fundamentals

Corrosion in aqueous systems is electrochemical. It always involves at least one oxidation reaction (the oxidation of metal to its ionic state), and at least one reduction reaction, with a flow of electrons through the metal. The site at which oxidation (corrosion) occurs is called the anode, and where reduction occurs, the cathode. As shown schematically in Figure 1.1, electrons released by the oxidation reaction flow through the metal from the anode to the cathode, where they are consumed in the reduction reaction. It is important to note that oxidation, reduction, and the flow of electrons must occur for corrosion to proceed. If either the oxidation reaction or the reduction reaction is stopped, or if the flow of electrons is interrupted, the entire corrosion process stops.

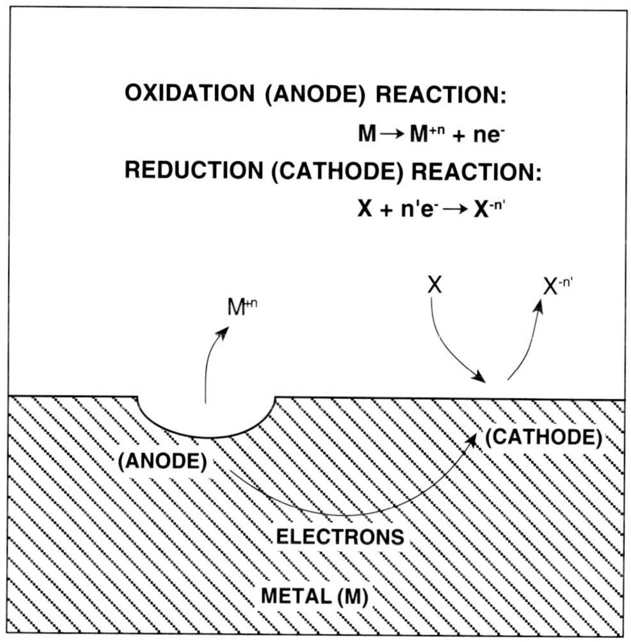

FIGURE 1.1
Idealized view of corrosion process, showing oxidation (corrosion) of metal "M" at the anode and reduction of some species "X" at the cathode. In neutral waters, it is mostly oxygen that is reduced at the cathode.

Figure 1.1 is actually a gross simplification of this process. In fact, several different reactions can occur at the anode or the cathode (including, for instance, one or more *reduction* reactions at the anode in addition to the oxidation reaction). What is shown in Figure 1.1 is the net reaction.

The anodic or cathodic reactions may be rate limited by availability of reactants, temperature, concentration of reaction products, and/or availability of essential catalysts. The reactions seldom proceed at theoretical rates because generally one or more of these factors is not optimum. The slowest reaction determines the rate of all reactions.

For example, if oxygen reduction is the dominant cathodic reaction, but oxygen cannot get to the metal surface because of the presence of a surface film of hydrogen, the anodic reaction with its accompanying release of electrons will be slowed, no matter how favorable the thermodynamics for that reaction. In neutral aqueous systems, it is usually the cathodic reaction that is rate limiting.

Polarization and Depolarization

When one reaction occurs at a different rate than the opposite electrode reaction, an electronic imbalance exists within the metal. In the typical case just described, the release of electrons at the anode is not matched by a corresponding consumption of electrons at the cathode. When this happens, the cathode is said to be "polarized." Removing the hydrogen film from the cathode allows the cathodic reaction to proceed. Thus, the cathode is said to have been "depolarized."

Induced Potentials and Cathodic Protection

Any applied potential can cause an electron imbalance in a piece of metal. Suppose the poles of a battery are connected to different areas on a metal surface. The area in contact with the positive pole of the battery will be polarized "positive," or will *tend* to be an anode and corrode. The area in contact with the negative pole of the battery will be polarized "negative," or will *tend* to be a cathode and not corrode. Thus, an externally applied, or "induced," potential can affect the location at which corrosion either occurs or does not occur on a metal surface.

The word "tend" is stressed for good reason. Suppose the area connected to the battery's negative pole has other factors that tend to make it anodic (i.e., that tend to make it corrode), and the battery is weak so that it only partly overcomes this anodic tendency.

In this case, that area is still an anode and probably will still corrode, albeit at a lower rate. By canceling out *some* of the anodic tendency, however, the battery will reduce the rate of corrosion that occurs at this anode.

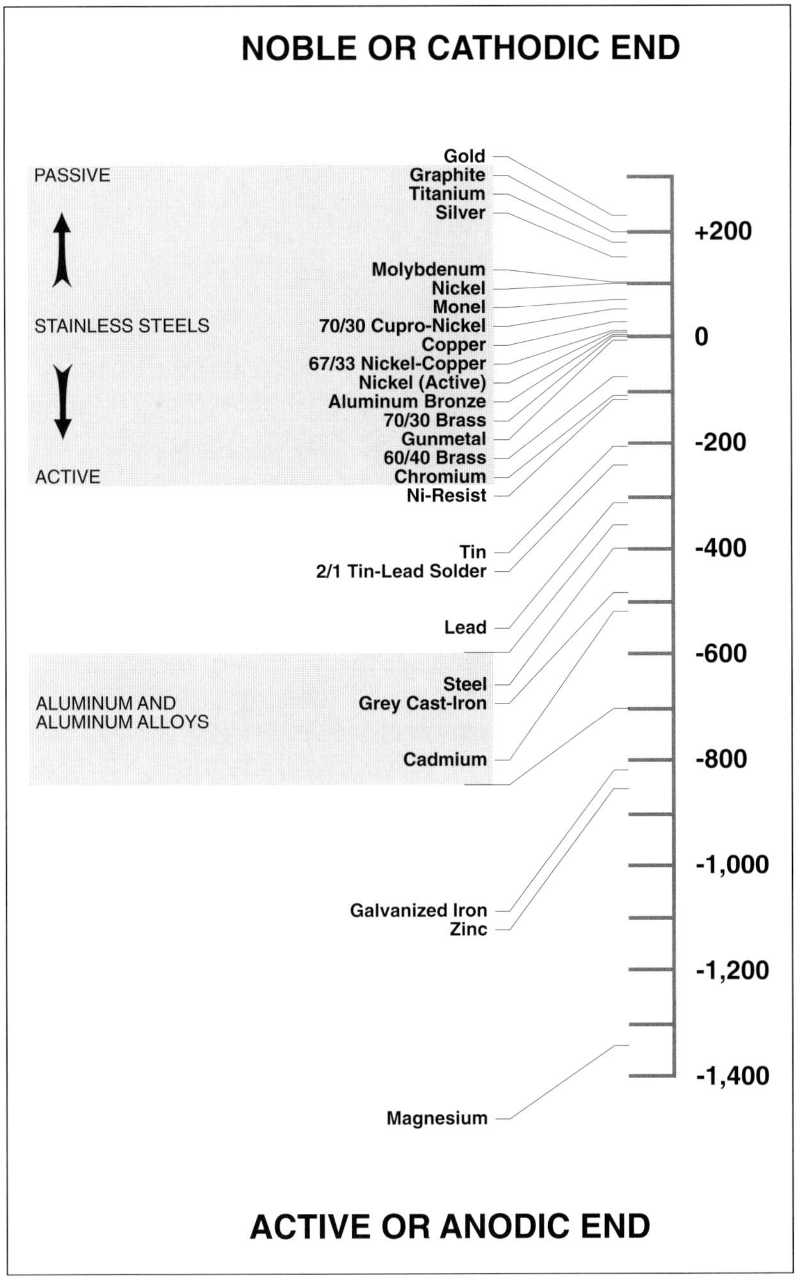

FIGURE 1.2
"Practical Galvanic Series of Metals and Alloys" - This is a composite galvanic series from a variety of sources and is not necessarily representative of any one particular environment. (After Butler and Ison[1])

This concept of applied potential has practical applications, especially in what is called "cathodic protection." In the case above, if the battery is sufficiently powerful, it will force the area in contact with its negative pole to be "cathodic," no matter how strongly other forces are working to make that area corrode.

The price for this "cathodic protection," however, is that the area in contact with the battery's positive pole will be forced to be anodic and is likely to corrode at a high rate.

However, suppose that the area in contact with the battery's positive pole is made into something that can pass electrons, but that will not corrode even when anodically polarized. In this case, the battery is doing the work of protecting a portion of the metal area, and the rest of the area is not suffering as a result.

In practice, an "inert" anode is suspended in the water, soil, or other electrolyte and connected to the positive pole of a power supply. The negative pole is then connected to the pipeline or other part to be protected.

Galvanic Corrosion

If a piece of metal is immersed in an electrolyte and connected through a voltmeter to a standard "reference" electrode also immersed in the electrolyte, a naturally occurring "galvanic" electrical potential will be observed between the metal and the reference electrode. If the metal is anodic to the reference electrode, the potential is stated in negative values, the implication being that the metal is tending to "corrode" more than the reference electrode. Then, if a similar experiment is run with a different type of metal, there will generally be a different galvanic potential between that metal and the same reference electrode.

In fact, the relative galvanic potentials measured in this experiment would tell exactly what would happen if the reference electrode was left out and the two different pieces of metal in the electrolyte were simply connected. They would exhibit a galvanic potential equal to the difference between their respective galvanic potentials relative to a reference electrode. One metal will act as an anode and will tend to corrode. The other will act as a cathode and will tend not to corrode.

By determining the galvanic potential of many different metals and alloys and listing these in order of relative galvanic potential, a "galvanic series" can be created for that particular environment (Figure 1.2).[1] The word *metal* implies a pure metal, such as iron or copper, whereas *alloy* denotes a mixture of metals or metals and functional elements intentionally added. Examples are steel (iron + carbon) and brass (copper + zinc). Alloys far apart on the list will create strong galvanic cells when coupled. The more "noble" (more electropositive) metal is cathodically protected at the expense of the "more active" (more electronegative) one. Coupling alloys that are close together on this electromotive series usually will not result in such galvanic corrosion.

It is important to remember that any galvanic series is specific for one particular environment (i.e., chemistry and temperature). The series for seawater at room temperature is widely published and is often used to explain galvanic corrosion problems. In freshwater, however, this series may not be applicable.

One example is the relationship between iron and zinc. In water at room temperature, the zinc is strongly anodic to iron and steel, which explains the basis for galvanized (zinc-coated) steel. At temperatures above approximately 140°F (60°C), depending upon water chemistry, their roles reverse, and steel will corrode preferentially to protect the zinc.

Corrosion cells in near-neutral aqueous systems are generally rate limited by the cathode reaction. This means that although there is adequate potential to drive the corrosion reaction faster, only a limited number of electrons will be released from a given area of cathode surface per unit time. However, if the cathode area is doubled, twice as many electrons will flow in the same amount of time, meaning dissolution (corrosion) at the anode can proceed twice as fast. Thus, it is important to consider not only the relative potential between the anode and cathode, but also the relative areas of the electrodes.

To illustrate this point, consider the old plumbers' rule of thumb: "You can put a brass valve in a steel line, but don't put an iron valve in a copper line." In most waters, brass and copper are cathodic to (i.e., higher on the electromotive series than) iron and steel. In the first case, the anode (steel pipeline) has a much larger area than the cathode (brass valve) so the relative area ratio inhibits corrosion. In the second case, the large copper surface will probably cause the smaller iron valve to corrode very rapidly.

Pitting and Crevice Corrosion

Most people see corrosion in their daily lives as rusting of steel parts such as auto bodies, bridges, etc. This corrosion is often fairly uniform and predictable. Rust (iron oxides) generally covers the corrosion in thin sheets or layers, thus the term "layered rusting." Similarly, uniform corrosion generally occurs on copper or brass with greenish corrosion products (oxides or sulfates of copper) or a thin, blackish film (copper sulfide). While corrosion in such cases is "general" (i.e., all over), it still occurs according to the anode/cathode description above. The electrodes continually exchange roles so the net effect is uniform corrosion.

Sometimes the conditions that create anodic behavior are very localized. These conditions may be an accumulation of corrosive salts, a small deposit on the metal surface that makes conditions different from those on the surrounding surface, or a crevice formed between two parts. If the conditions that form in such local sites tend to make the metal at that site more anodic than the surrounding metal surface, localized corrosion is likely to occur. Because such "occluded cells" are usually small when compared to the total metal surface, the relative area effect tends to cause rapid corrosion penetration.

When localized corrosion appears on an open surface and the aspect ratio is large (i.e., depth is roughly equal to or greater than the width of the corroded area), the corrosion is described as "pitting." When the corrosion occurs in a crevice and has a relatively small aspect ratio, the corrosion is called "crevice corrosion."

Passivation of Metals and Alloys

Steel and copper alloys are examples of metals for which the corrosion products that form on the surface usually offer only limited protection against further corrosion. Such metals most often exhibit general or uniform corrosion. In addition, steel and copper are both highly prone to direct oxidation in the presence of oxygen. For this reason, and because occluded cells such as deposits and crevices tend to be lower in oxygen than the surrounding surfaces, steel and copper alloys would be expected to corrode less under deposits and in crevices than on surrounding surfaces. It is not surprising, therefore, to find full metal thickness under nuts and washers on otherwise badly corroded steel assemblies.

Certain other metals and alloys (most notably stainless steels, nickel, and titanium) tend to form very tenacious and protective oxide films on the surface, which make them highly resistant to general corrosion. An exception to this resistance occurs in reducing environments, which can dissolve the oxide film. This oxide film is usually only several angstrom units thick, and thus invisible to the eye. The formation of such a film is called "passivation."

Contact with an oxygenated solution tends to reinforce the oxide film on these alloys. Conversely, absence of oxygen, or absence of an oxidizing condition, can lead to breakdown of the film. Therefore, so-called "passive" alloys are particularly susceptible to pitting and crevice corrosion in reducing environments. Halide ions, and especially chloride ions, aid the local breakdown of oxide films. Occluded cells that contain significant levels of chloride salts are particularly destructive. Certain alloying elements—notably molybdenum and tungsten—help to stabilize the passive film in the presence of chlorides and/or slightly reducing conditions, and are added to enhance resistance of passive alloys to chlorides and to localized corrosion in general.

Another very special case exists regarding stainless steels and oxidizing chloride salts. As stated previously, oxidizing conditions generally favor stabilization of the passive film, while chlorides tend to disrupt this film locally. Highly oxidizing chloride salts, such as ferric chloride ($FeCl_3 \cdot 6H_20$) and manganic chloride ($MnCl_3$), are uniquely potent "pitters" of stainless steels. Likewise, nonoxidizing chloride salts, such as NaCl, are also made more destructive by addition of highly oxidizing permanganate ions. The combination of ferric chloride and permanganate is probably the most destructive of all.

Basics of Microbiology and MIC

The microorganisms of interest in MIC are mostly bacteria and fungi. Algae deposits on metal surfaces may harbor bacteria and fungi. However, as that is their only perceived role at this time, algae will not be discussed further here.

A discussion of microorganisms and what they will and will not do is hampered by the fact that most microbiologists have traditionally done their work in the laboratory, using pure strains of the organism in question. The problem with this type of study is that organisms frequently behave differently in their natural environment than they do in the laboratory and they generally coexist in their natural environment in symbiotic relationships with other organisms.

Therefore, conclusions based on laboratory studies with pure cultures are misleading. Lab cultures are usually grown on rich media containing organic extracts that do not occur in the natural habitat of bacteria and fungi. Organisms in the environment usually compete for very low levels of nutrients. In the laboratory, the microbiologist is concerned with optimum growth conditions, while in the real world, organisms are usually doing what is necessary to survive in a hostile environment.

In order to influence either the initiation or the rate of corrosion in the field, microorganisms usually must become intimately associated with the corroding surface. In most cases, they become attached to the metal surface either in the form of a thin, distributed film or a discrete biodeposit. The thin film, or biofilm, is most prevalent in open systems exposed to flowing seawater, although the same can also occur in open freshwater systems. Such thin films start to form within the first 2 to 4 hours of immersion, but often take weeks to become mature (see next section on Biofilm Formation). These films will usually be spotty rather than continuous in nature, but will nevertheless cover a large proportion of the exposed metal surface.

In contrast to the distributed films are discrete biodeposits. These biodeposits may be up to several centimeters in diameter, but will usually cover only a small percentage of the total exposed metal surface. It is important to recognize that these two different distributions exist, and that they produce different influences on corrosion.

The organisms in the films and deposits discussed above will generally have a large effect on the chemistry of the environment within the deposit—that is, at the metal/film or the metal/deposit interface—without having any measurable effect on the bulk electrolyte properties. Occasionally, however, the organisms will be concentrated enough in the environment to influence corrosion by changing the bulk chemistry. This is sometimes the case in anaerobic soil environments, where the organisms do not need to form either a film or a deposit in order to influence corrosion.

That the taxonomy (identification and naming) of microorganisms is an inexact science is another frustrating fact for those dealing with MIC. Most microbiological laboratories are equipped to identify functional groups of organisms rather than specific strains. Most identification techniques are designed to find only certain types of organisms, while completely missing other types. The tendency is to identify the organisms that are easy to grow in the laboratory rather than the organisms prevalent in the field. This is particularly true of routine microbiological analyses by many chemical service companies, which, although proposing to be very specific, are often based on only the crudest of analytical techniques.

Bacteria can exist in several different metabolic states. Those that are actively respiring, consuming nutrients, and proliferating are said to be in a "growth" stage. Those that are simply existing, not growing because of unfavorable conditions, are said to be in a "resting" state. Some strains, when faced with unacceptable surroundings, form spores that can survive extremes of temperature and long periods without moisture or nutrients, yet produce actively growing cells quickly when conditions again become acceptable. The latter two states may appear, to the casual observer, to be like death, but the organisms are far from dead. Cells that actually die are usually consumed rapidly by other organisms or enzymes. When looking at an environmental sample under a microscope, therefore, it should be assumed that most or all of the cell forms observed were alive or capable of life at the time the sample was taken.

Classification of Microorganisms

Microorganisms are first categorized according to oxygen tolerance. There are:

- Strict (or obligate) anaerobes, which will not function in the presence of oxygen;
- Aerobes, which require oxygen in their metabolism;
- Facultative anaerobes, which can function either in the absence or presence of oxygen; and
- Microaerophiles, which use oxygen but prefer low levels (often referred to as "aerotolerant" microorganisms).

Strictly anaerobic environments are quite rare in nature, while strict anaerobes are commonly found flourishing within anaerobic microenvironments in highly aerated systems.

Another way of classifying organisms is according to their metabolism:

- The compounds from which they obtain their carbon for growth and reproduction (i.e., "nutrients"),
- The chemistry by which they obtain energy or perform respiration (i.e., "terminal electron acceptors"), and
- The elements they accumulate as a result of these processes.

Some of the classifications of particular interest in MIC are:

- Sulfate-reducing bacteria (SRB),
- Sulfur/sulfide-oxidizing bacteria,
- Acid-producing bacteria and fungi,
- Iron-oxidizing bacteria,
- Manganese-fixing bacteria
- Acetate-oxidizing bacteria, and
- Acetate-producing bacteria.

A third way of classifying bacteria is by shape. Examples are:

- "Vibrio," for comma-shaped cells,
- "Bacillus," for rod-shaped cells,
- "Coccus," for round cells, and
- "Myces," for fungi-like cells, etc. (See Figures 9.25-9.34 in Catalog of Color Photographs of MIC.)

These shapes are predictable when organisms are grown under well-defined laboratory conditions. In natural environments, however, shape is often determined by growth conditions rather than pedigree.

The most specific classification of organisms is by name: genus, strain, and, in some cases, source. For example, a description of one SRB might be "*Desulfovibrio desulfuricans*, Norway."

Bacteria Commonly Associated with MIC

SRB

All SRB are anaerobes. For the most part, they require a complete absence of oxygen and a highly reduced environment to function efficiently. Nonetheless, they circulate (probably in a resting state) in aerated waters, including those treated with chlorine and other oxidizers, until they find a site to their liking. There is also a growing body of evidence that some SRB strains can tolerate low levels of oxygen.

SRB are usually lumped into two nutrient (organic carbon) categories: Those that can use lactate and those that cannot. The latter generally use acetate and are difficult to grow in the laboratory on any medium. Lactate, acetate, and other short-chain fatty acids usable by SRB do not occur naturally in the environment. Therefore, these organisms depend on other organisms to produce such compounds. SRB reduce sulfate to sulfide, which usually shows up as hydrogen sulfide or, if iron is available, as black ferrous sulfide. In the absence of sulfate, some strains can function as fermenters and use organic compounds such as pyruvate to produce acetate, hydrogen, and carbon dioxide. Many SRB strains also contain *hydrogenase* enzymes, which allow them to consume hydrogen.

Most common strains of SRB grow best at temperatures from 25° to 35°C. A few thermophilic strains capable of functioning efficiently at more than 60°C have been reported. It is a general rule of microbiology that a given strain of organism has a narrow temperature band in which it functions well, although different strains may function over widely differing temperatures. However, there is some evidence that certain organisms, especially certain SRB, grow well at high temperatures (around 100°C) under high pressures (e.g., 2,500 to 4,500 psi), but can also grow at temperatures closer to 35°C at atmospheric pressure.

This evidence is sometimes used to explain why water samples from deep wells where temperatures are typically over 100°C show signs of SRB activity. However, others have proposed that two different types of SRB are present, one group capable of growing at more than 100°C and the other at 25 to 35°C.

Tests for the presence of SRB have traditionally involved growing the organisms on laboratory media, quite unlike the natural environment in which they were found. These laboratory media will only grow certain strains of SRB, and even then some samples require a long "lag time" before the organisms will adapt to the new growth conditions. As a result, misleading information has been obtained regarding the presence or absence of SRB in field samples.

Newer methods that do not require the SRB to grow to be detected have been developed. These methods are not as sensitive as the old culturing techniques but are useful in monitoring "problem" systems in which numbers are relatively high.

SRB have been implicated in the corrosion of cast iron and steel, 300 series stainless steels, copper-nickel alloys, and high nickel-molybdenum Alloy B. They are almost always present at corrosion sites because they are in soils and surface water streams and waterside deposits in general. Their mere presence, however, does not mean they are causing corrosion. The key symptom that usually indicates their involvement in the corrosion process is localized corrosion filled with black sulfide corrosion products.

While significant corrosion by pure SRB strains has been observed in the laboratory, in their natural environment these organisms rely heavily on other organisms to provide not only essential nutrients, but also the necessary microanaerobic sites for their growth.

Sulfur/Sulfide-Oxidizing Bacteria

This broad family of aerobic bacteria derives energy from the oxidation of sulfide to elemental sulfur or sulfate, or by oxidation of sulfur to sulfate. Those aerobic bacteria that produce elemental sulfur as their endpoint probably do not contribute directly to corrosion (elemental sulfur is not corrosive at near-ambient temperatures), but can form bulky deposits with anaerobic zones beneath, suitable for growth of SRB.

Other types of aerobes, however, may oxidize the sulfur to sulfuric acid, with pH values as low as 1.0 reported. These *Thiobacillus* strains are most commonly found in mineral deposits, and are largely responsible for acid mine drainage, which has become an environmental concern. They proliferate inside sewer lines and can cause rapid deterioration of concrete mains and the reinforcing steel therein. They are also found on surfaces of stone buildings and statues and probably account for much of the accelerated damage commonly attributed to acid rain (although the sulfur oxides in the acid rain may be their essential intermediate).

Where *Thiobacillus* bacteria are associated with corrosion, they are almost always accompanied by SRB. Thus, both types of organisms are able to draw energy from a synergistic sulfur cycle. The fact that two such different organisms—one a strict anaerobe that prefers neutral pH, and the other an aerobe that produces and thrives in an acid environment—can coexist demonstrates that individual organisms are able to form their own microenvironment within an otherwise hostile larger world.

Iron/Manganese-Oxidizing Bacteria

Bacteria that derive energy from the oxidation of Fe^{++} to Fe^{+++} are commonly reported in deposits associated with MIC. They are almost always observed in tubercles (discrete hemispherical mounds) over pits on steel surfaces. The most common iron oxidizers are found in the environment in long protein sheaths or filaments, much like the beans in a string bean pod.

While the cells themselves are rather indistinctive in appearance, these long filaments are readily seen under the microscope and are not likely to be confused with other life forms. (See Catalog of Color Photographs of MIC, Figures 9.33 and 9.34.) The observation that filamentous iron bacteria are "omnipresent" in tubercles might be, therefore, more a matter of their easy detection than of their relative abundance.

An intriguing type of iron oxidizers is the *Gallionella* bacterium, which has been blamed for numerous cases of corrosion of stainless steels. It was previously believed that *Gallionella* simply caused bulky deposits that plugged water lines. More recently, however, they have been found in several cases in which high levels of iron, manganese, and chlorides are present in the deposits. The resulting ferric-manganic chloride is a potent pitting agent for stainless steels.

Besides the iron/manganese *oxidizers*, there are organisms that simply *accumulate* iron or manganese. Such organisms are believed to be responsible for the manganese nodules found on the ocean floor. The accumulation of manganese in biofilms is blamed for several cases of corrosion of stainless steels and other ferrous alloys in water systems treated with chlorine or chlorine/bromine compounds. It is likely that the organisms' only role in such cases is to form a biofilm rich in manganese. The hypochlorous ion then reacts with the manganese to form permanganic chloride compounds, which cause distinctive subsurface pitting and tunneling corrosion in stainless steels. (Figures 9.11-9.14, Catalog of Color Photographs of MIC).

Aerobic Slime Formers

Aerobic slime formers are a diverse group of aerobic bacteria. They are important to corrosion mainly because they produce extracellular polymers that make up what is commonly referred to as "slime." This polymer is actually a sophisticated network of sticky strands that bind the cells to the surface and control what permeates through the deposit. The stickiness traps all sorts of particulates that might be floating by, which, in dirty water, can result in the impression that the deposit or mound is an inorganic collection of mud and debris. The slime formers and the sticky polymers that they produce make up the bulk of the distributed slime film (called a "primary film") that forms on all materials immersed in seawater.

Slime formers can be efficient "scrubbers" of oxygen, thus preventing oxygen from reaching the underlying surface. This creates an ideal site for SRB growth. Various types of enzymes, intended to help the cells grow and proliferate, are often found within the polymer mass, but outside the bacterial cells. Some of these enzymes are capable of intercepting and breaking down toxic substances (such as biocides) and converting them to nutrients for the cells.

Tubercles, though attributed by some to filamentous iron bacteria, usually contain far greater numbers of aerobic slime formers. The word *tubercles* usually refers to the relatively hard, frangible mounds found on steel surfaces. Softer mounds, similar to tubercles but lower in iron content, are also found on stainless steels and other metal surfaces, usually in conjunction with localized MIC. These, too, typically contain high numbers of aerobic bacteria, either *Gallionella* or slime formers (Figures 9.1, 9.2, and 9.15, Catalog of Color Photographs of MIC).

The term *high numbers* is relative. A microbiologist considers 10^6 cells per mL or per gm in an environmental sample to represent high numbers. However, these organisms make up only a minuscule portion of the overall mass. Biomounds, whether crusty tubercles on steel surfaces or the softer mounds on other metals, typically analyze approximately 10 percent by weight organic matter, most of that being extracellular polymers.

Methane Producers

Only in recent years have methane-producing bacteria (methanogens) been added to the list of organisms believed responsible for corrosion. Like many SRB, methanogens consume hydrogen and thus are capable of performing cathodic depolarization. While they normally consume hydrogen and carbon dioxide to produce methane, in low-nutrient situations these strict anaerobes will become fermenters and consume acetate instead. In natural environments, methanogens and SRB frequently coexist in a symbiotic relationship: SRB producing hydrogen, CO_2 and acetate by fermentation, and methanogens consuming these compounds—a necessary step for fermentation to proceed.

The case for facilitation of corrosion by methanogens still needs to be strengthened, but methanogens are as common in the environment as SRB and are just as likely to be a problem. The reason they have not been implicated before now is most likely because they do not produce distinctive, solid byproducts, and because they are not commonly included in MIC-related microbiological analyses.

Organic Acid-Producing Bacteria

Various anaerobic bacteria such as *Clostridium* are capable of producing organic acids. Unlike SRB, these bacteria are not usually found in aerated macroenvironments such as open, recirculating water systems. However, they are a problem in gas transmission lines and could be a problem in closed water systems that become anaerobic.

Acid-Producing Fungi

Certain fungi are also capable of producing organic acids and have been blamed for corrosion of steel and aluminum, as in the highly publicized corrosion failures of aluminum aircraft fuel tanks. In addition, fungi may produce anaerobic sites for SRB and can produce metabolic byproducts that are useful to various bacteria.

References

1. M.A. Butler, R.C.K. Ison, *Corrosion and Its Prevention in Waters* (New York, NY: Van Nostrand Reinhold Publishing Corp., 1966).

Biofilm Formation

By G.G. Geesey

In aquatic environments, microorganisms may be suspended freely in the bulk water (planktonic existence) or attached to an immobile substratum or surface (sessile existence). The microorganisms may exist as solitary individuals or in colonies that contain from a few to more than a million individuals. Complex assemblages of various species may occur within both planktonic and sessile microbial populations. The environmental conditions largely dictate whether the microorganisms will exist in a planktonic or sessile state.

Sessile microorganisms do not attach directly to the substratum surface, but rather to a thin layer of organic matter (the "conditioning film") adsorbed on the surface (Figure 1.3, Stages 1 and 2). As microbes attach to and replicate on the substratum, a biofilm is formed over the surface. The biofilm is composed of immobilized cells and their extracellular polymeric substances.

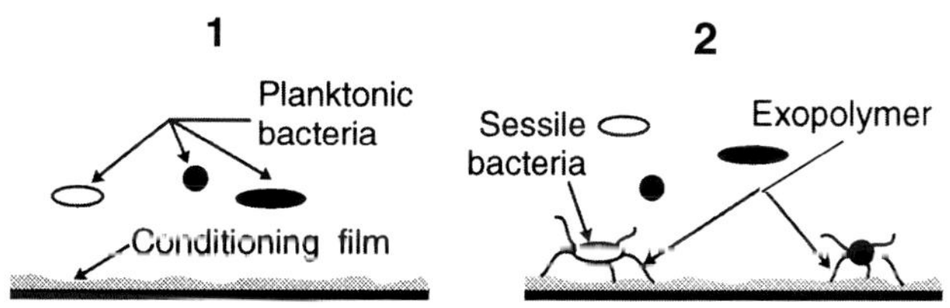

FIGURE 1.3

Biofilm Development

Stage 1: Conditioning film accumulates on submerged surface.
Stage 2: Planktonic bacteria from the bulk water colonize the surface and begin a sessile existence by excreting exopolymer that anchors the cells to the surface.

The characteristics of a biofilm may change with time. During the early stages of development, a biofilm is composed of the pioneering microbial species, which are distributed as individual cells in a heterogeneous manner over the surface. Within a matter of minutes, some of the attached species produce adhesive exopolymers that encapsulate the cells and extend from the cell surface to the substratum and into the bulk fluid (Figure 1.3, Stage 2). The adhesive exopolymers restrict the dissemination of microbial cells as they replicate on the surface (Figure 1.3, Stage 3). At this stage of development, the biofilm is less than 10 μm in thickness and exists as a discontinuous matrix of exopolymers interspersed with cells.

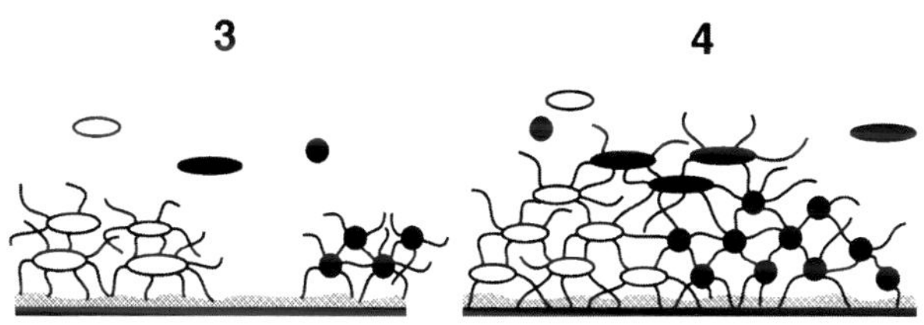

FIGURE 1.3

Biofilm Development

Stage 3: Different species of sessile bacteria replicate on the metal surface.
Stage 4: Microcolonies of different species continue to grow and eventually establish close relationships with each other on the surface. The biofilm increases in thickness. Conditions at the base of the biofilm change.

As the immobilized cells continue to replicate and excrete more exopolymer material, the biofilm forms a confluent blanket of increasing thickness over the surface (Figure 1.3, Stage 4). Over time, species of planktonic bacteria and nonliving particles become entrained in the biofilm and contribute to a growing community of increasing complexity. At this stage, the mature biofilm may be visibly evident. Its morphology and consistency vary depending on the types of microorganisms present and the conditions in the surrounding bulk liquid. The time it takes to achieve this stage may vary from a few days to several weeks.

As the biofilm increases in thickness, diffusion of dissolved gases and other nutrients from the bulk liquid to the substratum becomes impeded. Conditions become inhospitable to some of those microorganisms at the base of the biofilm, and eventually many of these cells die. As the foundation of the biofilm weakens, shear stress from the flowing liquid causes sloughing of cell aggregations, and localized areas of bare surface are exposed to the bulk liquid (Figure 1.3, Stage 5). The exposed areas are subsequently recolonized and new microorganisms and their exopolymers are woven into the fabric of the existing biofilm (Figure 1.3, Stage 6). This phenomenon of biofilm instability occurs even when the physical conditions in the bulk liquid remain constant. Thus, biofilms are constantly in a state of flux.

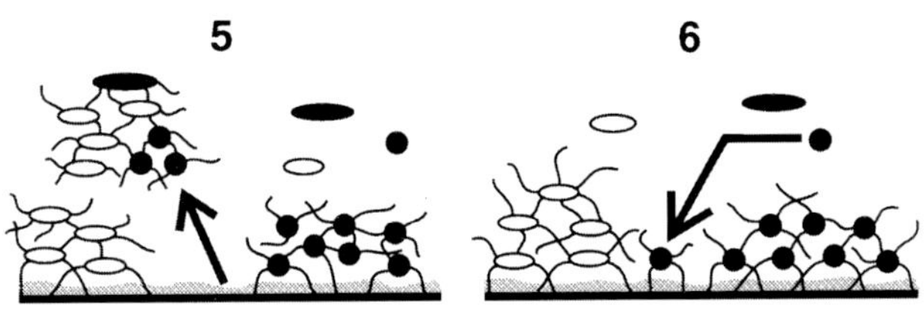

FIGURE 1.3

Biofilm Development

Stage 5: Portions of the biofilm slough away from the surface.

Stage 6: The exposed areas of surface are recolonized by planktonic bacteria or sessile bacteria adjacent to the exposed areas.

Effect of Operating Conditions on MIC

Biocorrosion problems most often occur in new systems when they are first wetted. When the problem occurs in older systems, it is almost always a result of changes such as new sources or quality of water, new materials of construction, new operating procedures (e.g., water now left in system during shutdowns, whereas it used to be drained), or new operating conditions (especially temperature). Some of the operating parameters known to, or suspected of, having an effect on MIC are temperature, pressure, flow velocity, pH, oxygen level, and cleanliness.

Temperature

All microorganisms have an optimum temperature range for growth. Observation of the water or surface temperatures at which corrosion mounds or tubercles do or do not grow may offer important clues as to how effective slight temperature changes may be. Cooling water systems often allow little latitude in temperature, but other water systems may permit increasing or decreasing temperatures significantly. In one cooling water system, an energy conservation effort placed automatic flow control on heat exchangers to maintain maximum permissible outlet temperature. The result was an average increase of approximately 5°C in the water temperature, with a corresponding doubling or tripling of corrosion and fouling rates. The electricity saved on pumping was offset by increased maintenance costs.

The normal expectation is that any corrosion problem is increased by increasing temperatures. With MIC, this is not necessarily so. In one case of crevice corrosion of stainless steel at approximately 35°C, the problem disappeared if the temperature was increased to 70°C. Unfortunately, this was of no practical help in that particular system, but it did help to prove that the problem was microbiological in nature.

Flow Velocity

Flow velocity has little long-term effect on the ability of cells to attach to surfaces. Once attachment takes place, however, flow affects the nature of the biofilm that forms. It has been observed that low-velocity biofilms tend to be very bulky and easily disturbed, while films that form at higher velocities are much denser and thinner—and more tenacious.

As a rule, flow velocities above 5 ft/s (1.5 m/s) are recommended in water systems to minimize settling out of solids. Such velocities will not prevent surface colonization in systems prone to biofouling, however. Stagnant conditions, even for short periods of time, generally result in problems.

Increasing velocity to discourage biological attachment is not feasible and generally promotes erosion-corrosion or velocity-accelerated corrosion of the particular metal being used. Copper, for instance, suffers erosion-corrosion above ~ 5 ft/s (1.5 m/s) at 20°C.

pH

Bulk water pH can have a significant effect on the microorganisms. Growth of common strains of sulfate-reducing bacteria (SRB), for example, slows above pH 11 and is completely stifled at pH 12.5. Some researchers have speculated that this is why cathodic protection is effective against these microbes, in that it generates free alkalinity at the metal surface.

Oxygen Level

Many bacteria require oxygen for growth. It was once thought that problems with anaerobes, such as SRB, could be alleviated by sparging the system with air. This has since been disproved. Because SRB depend on other organisms (often aerobes) for their nutrients, and they very effectively "hide" in aerated systems, aeration will not reduce the impact of SRB and, in some cases, exacerbates the problem.

There is reason to believe, however, that many biological problems would improve if the system were *completely* deaerated. The key is that *all* oxygen must be removed. Many aerobes and microaerobes can function adequately with as little as 50 ppb O_2, and facultative organisms, of course, simply convert to an anaerobic metabolism if oxygen is depleted. Practically speaking, removing dissolved oxygen from the system can affect MIC, but it is not likely to eliminate a severe problem.

Cleanliness

The "cleanliness" of a given water usually refers to the water's turbidity or the amount of suspended solids in that water. Settling of suspended solids enhances corrosion by creating occlusions and surfaces for microbial growth and activity. The organic and dissolved solids content of the water are also important. These factors may be significantly reduced by "cleaning up" the water.

Improving water quality is not necessarily a solution to MIC. In one system that was fed dirty river water, tower makeup was switched to clarified and filtered water from the same source. The water appeared noticeably cleaner, and the biofilms that formed from time to time were relatively unnoticeable because they had little color. The mounds that grew over corrosion sites were significantly smaller because of the lack of trapped suspended solids. However, the bottom line was that biological control was just as difficult as before, and MIC was just as severe.

With respect to water cleanliness, one rule is that as long as *any* microorganisms can grow in the water, the potential for MIC exists. On the surfaces of piping and equipment, however, "cleanliness" is much more important. Anything that can be done to clean metal surfaces physically on a regular basis (i.e., to remove biofilms and deposits) will help prevent or minimize MIC.

In summary, any time the operating conditions in a water system are changed (including chemical treatment, which will be discussed later), extra attention should be paid to possible biological problems that may result. If a problem exists, consider changes in operating conditions first as a cost-effective means of alleviating the problem.

MIC in the Chemical Industry

By J.G. Stoecker

Microbiologically influenced corrosion (MIC) is a persistent problem in the chemical industry. However, it has taken a considerable amount of multidisciplinary work over the last decade to separate MIC from other forms of corrosion (such as crevice and under-deposit corrosion). The separation and identification of MIC as a specific problem prompted the Materials Technology Institute of the Chemical Process Industries, Inc. (MTI) to fund a study of the phenomenon. The results of this study are included in MTI Publication No. 13, *Microbiologically Influenced Corrosion: A State of the Art Review*, available from NACE International.[1] A review of this publication and others concerning case histories of MIC in the chemical industry reveals a number of areas in which the problems associated with MIC are most persistent.

Complications with the water used for hydrostatic testing and/or running in new processes and equipment are the most visible problems resulting from MIC in the chemical industry. This is because of the difficulties involved in explaining how new equipment can corrode even before chemicals are added to the process system. MIC is also a pervasive problem in cooling water systems. However, problems associated with MIC in cooling water systems do not receive much attention in the literature because specific MIC problems in these systems tend to become lost in the myriad of other water treatment problems. Case histories of MIC of underground pipe and vessels can also be found in the chemical industry literature. Recently, MIC problems associated with demineralized and even deionized waters have been cited. In general, MIC is a potential problem in any aqueous system in the chemical industry. An aqueous system is defined as any system that contains even trace amounts of water, e.g., water in fuel oil tanks.

This discussion addresses the most persistent MIC problem areas in the chemical industry, along with some of the most helpful prevention techniques. Areas addressed include hydrostatic testing, buried structures, and materials selection.

Hydrostatic Testing

Microbiological species found in water used for hydrostatic testing equipment, running in continuous processes before chemical startup and water batching in batch processes before chemical startup, have caused considerable damage and expense in the past.[2-4] In an effort to solve this expensive problem, guidelines for hydrostatic testing have been adopted by several companies. Where they have been followed to the letter, these guidelines have produced positive results. Unfortunately, use of these guidelines can be very expensive and not always completely quantifiable or justifiable. This usually leads to a cost-cutting exercise that produces a less expensive adaption of the guidelines. Adapted programs have produced mixed results.

The following discussion describes several hydrostatic testing guidelines and the problems associated with their use. Cost-cutting ideas that do not reduce the effectiveness and purpose of the hydrostatic testing guidelines are also presented. However, the use of commercial biocides is beyond the scope of this paper and, therefore, is not included. It should be noted that the disposal of biocides may pose significant difficulties.

Demineralized water is probably the first choice for use in hydrostatic testing.[4] The word "probably" is inserted in this statement because of a study involving microbiological corrosion in demineralized water.[5] The results of this study confirmed the importance of draining and drying the system at the earliest opportunity after hydrostatic testing. This must be included as part of the procedure if demineralized water is chosen for the hydrotest fluid.

Consider, however, the cost of hundreds of thousands of gallons of demineralized water required for a large process. Also consider the difficult task of drying a large process system after testing. Management may be very reluctant to spend the large sums associated with a program of this magnitude. The positive aspect of using demineralized water is that disposal usually is not a problem.

High-purity steam condensate with early draining and drying of the system after testing is the second choice for hydrostatic testing. Again, expense is an important factor. Finding a steam system large enough to tolerate sufficient bleed-off to test a large process system can be a problem, not to mention finding a system that is not full of treatment chemicals. Caution must be exercised in using condensate that contains treatment chemicals because these compounds may be corrosive to the process equipment being tested or may contaminate the product being produced.

Depending on the chemicals present in the condensate, disposal can be a problem. Because of these limitations, this testing program is generally harder to sell than the demineralized water test noted above.

The use of natural freshwater from wells, lakes, the city or plant water system, etc. is the least expensive alternative. Freshwater is the last choice for hydrostatic testing and produces the greatest risk of bacterial infection and microbiological corrosion. Even with the use of freshwater, the effect of bacterial infection can be reduced by immediately draining the system after testing, flushing the system with demineralized water or steam condensate and/or blowing or mopping the system dry within three to five days after testing.

Fortunately, there are some options to be considered in hydrostatic testing that can lessen the expense, the disposal problems, and the difficulty of selling a program to management.

Consider the following:

• The chemicals in the process to be tested. If it is determined that microbes will not survive in the chemicals, consider activating the system immediately after testing with any of the choices listed above, e.g., demineralized, high-purity steam condensate or natural freshwater.

• Reusing the hydrotest water and thereby reducing the volume and expense. Chlorine and other treatments can be used effectively to control microbes in a water reuse program. For this treatment to be effective, a periodic test must be performed to ensure that the water contains adequate biocide at all times. The pH, oxygen level, temperature, chloride concentration, etc. of the water must be considered if the system being tested contains 300 series stainless steel.

An uncontrolled chlorine treatment may result in pitting or chloride stress corrosion cracking (SCC) of these alloys. Chlorides produced during the treatment may cause damage to these alloys both during the hydrotest and during process operation. Chloride levels in the water during hydrostatic testing may not constitute a risk to these alloys at ambient temperature. However, even trace levels of chlorides left in the system from the hydrostatic test water may be a potential risk to these alloys under high-temperature process conditions. Disposal of the water will depend upon the level of free chlorine or other biocides in the water.

• Reusing the hydrotest water, as described above, but with an ozone treatment. Ozone has been used effectively in recent years to eliminate microbe activity. In addition, experience has shown that ozone at levels of 0.2 ppm has an excellent ability to penetrate dense microbiological deposits and kill the microbes. Ozone has a very short life, which eases water disposal problems. When required, ultraviolet light has been used effectively to destroy ozone and thereby eliminate the problems associated with water disposal. However, ozone should be used with caution because it can damage many materials, particularly plastics and elastomers.

• Consider pneumatic testing for the process system rather than hydrostatic testing. This is not a complete replacement for hydrostatic testing because pneumatic testing is primarily a leak test and not a strength-plus-leak test, as is hydrostatic testing. Pneumatic leak tests of the process system combined with strength-plus-leak hydrostatic tests of vessels and other components at the fabricator's facility can be used quite effectively in proving a system. On an individual basis, vessels and other process components can easily be dried after hydrotesting at the fabricator's shop. This eliminates the threat of microbiological corrosion, and allows freshwater, which presents little or no disposal problem, to be used. Safety during pneumatic testing is the major concern with the combined pneumatic/hydrostatic program. Safe practices and pressure limits must be strictly followed during pneumatic testing.

• Consider all sources of microbiological contamination when constructing a facility or fabricating process equipment. Hydrostatic test water in a fabricator's shop can be a source of microbiological contamination that can result in corrosion of the equipment being fabricated and even of the entire process system. Draining and drying immediately after hydrostatic testing should be included in all fabrication specifications. Placing vessels, pipes, and other process components on the ground in construction laydown yards, which are usually muddy, can also lead to microbiological contamination and damage. Good drainage of outdoor yards is extremely important because it can prevent ponding of rainwater and other foreign materials where bacteria can locate. The elimination of these ponds will reduce the opportunity of contamination of equipment that is stored in the yard.

Buried Structures

Corrosion of buried pipelines caused by sulfate-reducing bacteria (SRB) has been studied for almost a century. Quite by accident, industry has been protecting buried structures from bacterial damage through the use of cathodic protection. Cathodic protection produces a caustic or basic environment on the surface of the buried structure. This caustic or basic environment is not conducive to microbiological growth. Booth was one of the early investigators to present this finding.[6] It is recommended that the benefits and risks of cathodic protection be evaluated for each material and each application by a qualified expert in cathodic protection.

No dramatic improvements in the microbiological protection of buried structures have been observed over the years. Cathodic protection used with a protective coating system has continued to be the best defense against microbiological attack and other underground corrosive damage. Cathodic protection provides the caustic protection at holes or holidays that are sure to develop with time in the coating.

In addition to cathodic protection and coating, backfilling with limestone or other caustic materials is an added step that can be taken to protect buried structures from microbiological damage. Providing adequate drainage to produce a dry environment both above and below ground in the area of the buried structure will also reduce the risk of this type of damage.

Materials Selection

The corrosion engineer's answer to corrosion problems often includes an upgrade in the material of construction. Although most solutions to corrosion problems include a material upgrade as an option, this approach may not be satisfactory for MIC.

For example, an upgrade from AISI 304 (UNS S30400) to 316 SS (UNS 31600) does not always solve the problem. Kobrin reported corrosion of delta ferrite stringers in weld metal.[4] An upgrade from AISI 304 to 316 SS would not solve the problem in this case because AISI 316 SS can contain as much or more delta ferrite than AISI 304 SS. Kobrin also reported microbiological corrosion of nickel, nickel-copper alloy 400 (UNS N04400), and nickel-molybdenum alloy "B" heat exchanger tubes.[4] Although the alloy 400 and alloy "B" tubes were not pitted as severely as the nickel tubes, the higher alloys did not completely solve the MIC problem. It should be noted that the issue of selective phase attack in weld metals is not clear because some authors report attack of delta ferrite while others report attack of austenite. Therefore, the reader is advised to review the various references and judge accordingly.

In the past, copper was believed to be toxic to microbiological species. This has not turned out to be the case, as pointed out in the previous example and as reported elsewhere.[2] Personal discussions with representatives from various power companies concerning microbiological corrosion of copper and copper-nickel alloys also prove that copper is no deterrent to this type of damage. At this point, tantalum, titanium, and zirconium seem to be the only relatively common metals that are not subject to microbiological damage.[7]

Potential for MIC

Quantifying the potential for problems associated with MIC in an aqueous system can be a difficult task. An investigation conducted to develop a history of MIC for the water source under consideration is a good first step. This history should include information concerning all corrosion problems associated with the water source in the plant or facility in question and in other plants or facilities using the same source. All corrosion problems in the water system should be investigated because a MIC problem may not be identified as such.

A list of questions that focuses attention on subject areas critical in identifying MIC problems should be developed as part of the investigation. This list is especially helpful if the investigation is being conducted by telephone or mail survey. However, an on-site investigation is often more effective than telephone or mail surveys. The following list of questions concerning the water source is not meant to be all inclusive, but is designed to help the reader develop an approach to the investigation of MIC problems.

- Are there corrosion problems in this water?
- Are there investigative reports on corrosion problems?
- Are there temperature, pH, flow rate, etc. data on the water in which the corrosion problem occurred?
- Is the water chemically treated?
- Why is the water treated?
- Are there other people who have had corrosion problems with this water?

In addition to conducting an investigation, tests can be conducted using the water source under consideration to determine the potential for MIC problems. It is recommended that these tests be conducted to supplement historical data. The tests should include chemical and microbiological analysis of the water plus *in situ* exposure tests. The conditions of the coupon exposure or "bug pot" tests should match the environmental conditions of the proposed facility. Several procedures are available commercially to test for MIC and are discussed later in Chapter 8, Identification of MIC.

References

1. D. Pope, D.J. Duquette, A. Johannes, P. Wayner, *Microbiologically Influenced Corrosion: A State of the Art Review*, 2nd ed., publication no. 13 (St. Louis, MO: MTI, 1989).
2. J. Stoecker, "Guide for the Investigation of Microbiologically Induced Corrosion," *Materials Performance* 23, 8(1984): p. 48.
3. R. Tatnall, J. Stoecker, R. Schultz, G. Kobrin, A. Piluso, "Case Histories: Bacteria Induced Corrosion," *Materials Performance* 20, 8(1981): p. 41.
4. G. Kobrin, "Corrosion by Microbiological Organisms in Natural Waters," *Materials Performance* 15, 7(1976): p. 38.
5. J. Stoecker, D. Pope, "Study of Biological Corrosion in High Temperature Demineralized Water," *Materials Performance* 25, 6(1986): p. 51.

6. G. Booth, *Corrosion by Microbiological Organisms in Natural Waters* (London, UK: Mills and Boon, 1971).
7. D.H. Pope, D.J. Duquette, A.H. Johannes, P.C. Wayner, "Microbiologically Influenced Corrosion of Industrial Alloys," *Materials Performance* 23, 4(1984): p. 14.

MIC in the Power Industry

By A.A. Stein

Introduction

Corrosion is an ever-present degradation mechanism in wetted components and systems. There are many forms of corrosion in metals, i.e. pitting, stress corrosion, general corrosion, galvanic corrosion, etc., and these have been recognized and are continually being addressed by the power industry. Only recently has the power industry recognized that corrosion is influenced by the presence of microbes. The problem is the same for both fossil and nuclear plants, but its effect is greater in nuclear plants. This is because of the inherently more conservative design basis, operating philosophy, and standards governing nuclear plants compared to fossil power plants. Leaks or interrupted flow in critical systems are totally unacceptable in nuclear plants. Economically, the impact of microbiologically influenced corrosion (MIC) on nuclear plants is significant. Each day a nuclear power plant is shut down, lost generating capacity will cost the plant $750,000 to $1 million. Lost revenue from a shut-down fossil plant could be one quarter of this. Consequently, utilities operating nuclear plants are continually evaluating all forms of degradation in an effort to implement modifications that will improve plant reliability, reduce maintenance, and increase revenues.

Background

A system's first encounter with MIC usually occurs during its initial exposure to an aqueous environment, such as during hydrotest, wet lay-up, or moist soil. MIC is a corrosion process influenced by the presence of bacteria. Microbes require water for survival and a source of energy for growth. They can grow in fluids with pH values ranging from -1 to 10, where -1 is the most acidic concentration, and derive energy from organic or inorganic materials. Microbes can survive temperatures which range from -4 to 210°F (-20 to 99°C).

TABLE 3.1
Microorganisms Associated with Corrosion

Microorganism	Description
Thiobacillus	Sulfide and sulphur oxidizers—acid producers—tolerate extremely high acidic environments—aerobic
Gallionella	Iron oxidizers—prefer low-acid conditions
Desulfovibrio	Sulfate reducers—prefer neutral to alkaline conditions—strictly anaerobic
Pseudomonas	*Aspergillus*
Sphaerotilus	*Hormodendrum*

MIC affects structural materials such as plain carbon steel and stainless steels (SS), nickel base alloys, and copper alloys. The primary corrosion-causing bacteria are believed to be the sulfate-reducing (SRB) and iron-oxidizing bacteria, but many more species of bacteria that can promote corrosion have been identified (Table 3.1). These bacteria can form colonies beneath which corrosion can occur. If colonization can be prevented, MIC usually will not occur. The presence of nodules in SS systems is a good indicator of colonization and MIC. In carbon steel, the presence of nodules may be due to microbial or chemical corrosion mechanisms. At some plants, up to 200 different forms of bacteria have been identified. Pitting is the usual form of corrosion. Microbiological attack has occurred in service water systems, heat exchangers, fire protection systems, demineralized water storage tanks, and fluidized gas desulfurization systems.

Occurrence

Service Water Systems, Including Heat Exchangers

Service water systems are fabricated from polymer-lined or bare carbon steel, SS, or copper-nickel materials. In lined pipe, MIC can occur at areas of lining disbondment and deterioration such as field welds and the crevices created by socket welds. The bacteria can use the coloring agents in some lining systems as a food source, degrading linings and promoting further lining deterioration. MIC generally occurs in regions where the flow rate is or has been less than 5 ft/s. If biocide additions are not maintained, bacteria can enter and colonize and form nodules which provide protection to the colony. Once bacteria colonize, much higher levels of biocide are required and even these levels may not be effective because it is difficult for most biocides to penetrate a nodule.

Silt and debris that enter the system can settle in low-flow and stagnant regions, providing areas for bacteria to reside and promote corrosion. Heat exchangers are susceptible to MIC when in wet lay-up or operating under low-flow velocity, producing pits in tubing. Stagnancy in pipe systems can also result in corrosion due to MIC. For example, pitting corrosion rates can be so great as to produce through-wall pitting in SS pipe with 0.375-inch wall in less than six months.

In service water systems, the severity of MIC depends on a variety of factors such as material, system design, water, and water treatment. In all cases, once MIC is established and left unattended, through-wall leaks, loss of structural integrity, flow deficiencies, loss of heat transfer, and component failures can result. In the 300 series

TABLE 3.2
Occurrence of MIC in Power Plants

System/Component	Material	Evidence
1. Fire protection	Carbon steel	Clogging; through-wall pits at threaded connections
2. Heat-exchanger tubing	Aluminum brass 70/30 Cu Ni 90/10 Cu Ni	Pitting; sulfur deposits present with pitting
	Incoloy 800	SRB; organic acid producers; pitting
3. Demineralized water storage tank	Type 316 SS	Rust nodules; vertical rust stain at welds
4. Service water pipe	Type 316 SS weld	Through-wall pitting in welds
	Type 316L SS lined carbon steel	Through-wall pits in welds; pits at lining discontinuities; nodules attached to lining; clogging
	Coated carbon steel buried pipe	Pitting where lining damaged during backfill by sharp stones
5. Cooling tower	Galvanized steel; wood	Wood rot
6. Pump impeller	SS	Pitting
7. Spent-fuel pool	Type 304/316 SS	Microbiological bloom in water, reducing visibility

SS systems, pits caused by MIC are usually found in weld deposits and heat-affected zones. These pits usually have narrow openings at the inside diameter surface with larger, more cavernous or bulbous interiors. In carbon steel systems, the pitting is broader in width and randomly spread throughout the system. Copper-based alloys are also susceptible to MIC, especially by SRB.

Periodic cleaning and water treatment can significantly reduce the concern for MIC in heat exchanger tubing. Cleaning techniques include the injection of sponge balls that continually clean the tubes.

Once MIC has been detected, periodic nondestructive test (NDT) inspections should be performed. The inspection methods available include visual, radiographic, ultrasonic, and eddy current. Visual inspection alone is adequate to identify the locations in systems fabricated from SS where nodules are present. A volumetric inspection such as radiography can quantify the damage. Visual inspections are not an acceptable method of inspecting carbon steel systems. Once areas of MIC are identified, further NDT evaluations may be required to more fully quantify the amount of MIC and to assess component structural integrity. Flow monitoring cannot be used to assess MIC damage because MIC can be present with adequate flow and may not be present if flow is blocked.

Damage due to MIC has occurred on tube sheets, tubes, and water boxes. The materials affected include the copper base alloys (70/30 and 90/10 copper-nickel and aluminum brass), the 300 series SS, and carbon steels. The failure modes reported are pitting and dealloying—both typical MIC failure modes.

These results reveal that the copper alloys, long believed to be an effective tool to mitigate fouling due to clams, are not resistant to the bacteria which can influence corrosion. Some of the failures in the 300 series stainless components occurred in welds, but nonwelded stainless tubing has also failed.

MIC can be confused with macrofouling (e.g., Asiatic clams), which causes flow blockage or flow reduction. Some plants have reported a reduction in flow rates, an increased pressure drop, and a reduction in thermal performance as a result of fouling or corrosion product buildup, not MIC. Corrosion due to bacteria can be active under fouling deposits. Flow monitoring and pressure drops are not effective methods for predicting the presence of MIC. However, continuous high-velocity systems (>6 ft/s) can minimize the concern for MIC.

Valves

MIC does not appear to be a concern for valve components. The reasons are not clear but may be due to the thicker walls present in valve bodies, providing more corrosion allowance.

Fire Protection Systems

Fire protection systems are fabricated from carbon steel and typically contain untreated water under stagnant conditions for long periods. The water rapidly deaerates due to reaction of oxygen with the pipe wall surface. These deaerated conditions provide an environment conducive to anaerobic bacteria. MIC usually produces leaks at pipe joint connections or results in clogging of the pipe, which affects the ability to provide sufficient water. MIC control has been achieved by treating the water to pH 10, use of biocide, and deaeration.

Demineralized Storage Tanks and Spent-Fuel Pools

Demineralized water storage tanks have experienced corrosion due to bacteria. The bacteria can easily pass through the associated ion exchange systems. In one case, treating with 200 ppm hydrogen peroxide for 48 hours was effective in eliminating the problem.

Spent-fuel pools contain very high purity water to store (temporarily) nuclear fuel removed from reactors. During fuel transfer to and from the pool, hydraulic fluid can drip from the transfer equipment into the pool. This fluid provides nutrients and promotes the growth of otherwise low concentrations of bacteria and algae, which reduces water clarity. Regulations require that the pool be clear with no obstruction of vision. To restore water clarity, high dosages of hydrogen peroxide can be added to kill the bacteria.

Pumps

MIC has occurred in an auxiliary feedwater pump because flushing water remained stagnant for a six-month period. The fluid was treated sewage effluent. The level of sewage treatment is unknown, but typical sewage treatment plants treat sewage only to remove bacteria harmful to humans. Therefore, SRB and other corrosion-influencing bacteria may survive and pass through. When a chlorine water treatment was implemented and the pumps operated, the problem appeared to have been corrected.

In salt water service, MIC has occurred in a pump flange weld to the pipe connection and in the pump discharge head column welds down to the impeller. In both cases, the materials were 300 series SS. One was a type 316L casting; and the other, unidentified. The problems appeared to be due to improper lay-up, which included draining but not drying. Therefore, MIC can become a concern for pumps if bacteria are allowed to gain a foothold and establish colonies. The high velocities present during pump operation may not be sufficient to remove the colonies once they are established. Lay-up procedures need to be established, and if wet lay-up is used, a periodic biocide slug injection may be required to prevent the establishment of bacteria colonies.

MIC does not occur when biocide is injected before the pump. Because the injection points are usually relatively close to the pump, the biocidal action is most potent and the biocide does not undergo significant decay. The local injection point suggests the benefit of multiple injection points in a cooling water system to obtain maximum biocidal action. Coatings, welding, or upgrading of pump material are repair methods. None of these specifically address MIC and, therefore, may be temporary if MIC re-establishes itself.

Fluidized Gas Desulfurizer Systems

SRB are usually present in wet fluidized gas desulfurizer (FGD) systems and can eventually result in corrosion of the carbon steel and SS materials. To date, corrosion failures resulting from MIC have not been widely publicized in these systems, probably due to more aggressive environments being present. However, MIC has occurred in SS absorbers in nonwelded areas. High concentrations of bacteria are known to be present to such a degree that a black-brown slime can be readily observed. The major impact of bacteria in FGD systems relates to plant safety concerns due to generation of H_2S, and the impact on the commercial value of gypsum with bacteria present. Hydrogen sulfide is heavier than air and will sink to low points in the system.

The possibility of H_2S collecting in slurry tanks can be reduced by first venting the tank and draining the slurry. The presence of bacteria in gypsum should not be harmful, but for aesthetic reasons the gypsum can be sterilized. Biocides to control MIC in these systems may be used provided they are compatible with the gypsum being produced.

Bibliography

1. A.A. Stein, "Microbiologically Induced Corrosion, An Overview." Paper presented at the Third International Symposium of Environmental Degradation of Materials in Nuclear Power Reactors held in Traverse City, MI, in 1987.
2. *Proceedings of the EPRI Workshop on Microbe Induced Corrosion* (Palo Alto, CA: Electric Power Research Institute [EPRI], Research Reports Center, 1986).

CHAPTER 4

MIC in the Pulp and Paper Industry

By R.W. Lutey

Overview

Reports indicate that more than 30 percent of the maintenance costs in the pulp and paper industry are directly related to corrosion. Recent studies demonstrate that much of the corrosion occurring in pulp mills and paper machines is a result of the uncontrolled growth of microorganisms that can influence corrosion directly or establish conditions that contribute to accelerated electrochemical corrosion. Although a specific cost figure for losses from microbiologically influenced corrosion (MIC) is not available, it is safe to assume that this figure is substantial.

Materials of construction used in the pulp and paper industry are not significantly different from those used in other industrial processes that are subject to corrosion. However, the environment within the pulp and papermaking process is particularly conducive to the growth of the microorganisms that cause MIC. The prevention of other microbiological problems, i.e., slime deposits, plugging and fouling, odor formation, biodeterioration, etc., is a routine responsibility in efficient papermaking. However, routine procedures to prevent or mitigate MIC have yet to be established by the industry, in part because of a lack of understanding of the basic technology of MIC and the difficulty of recognizing MIC in the operating environment.

Considerable investigation of MIC in pulp mill and paper machine systems has been conducted, and significant progress has been made in developing procedures to prevent or mitigate existing MIC problems. This paper highlights some of that work and provides a basis for understanding what can be done to address the problem. As that understanding becomes more widespread, MIC control will become more routine within the industry.

Locations of Potential MIC

This section lists several locations in pulp mill, paper machine, and allied systems where MIC has been reported.

Raw Water Systems

MIC can occur in both ground water (well water) and surface water systems. Shallow freshwater wells can be contaminated with iron-oxidizing bacteria such as *Gallionella sp.*, sessile bacteria that colonize (adhere) on the inner surface of piping used to transport raw water to treatment plants. Infestations have been observed from the wellheads all the way through filtering and sedimentation pretreatment processes. Often, the bacteria oxidize the soluble iron in the raw water. The bacteria can also attack the steel pipe. This can result in the establishment of an anodic area that is perpetuated by the growth of the bacteria and can cause continued corrosion of the pipe by conventional electrochemical mechanisms. The result is considerable fouling by the residual corrosion byproducts or, in severe situations, actual pipe failure.

MIC in raw water systems that use surface water sources is often associated with the deposition of sludge and silt in low-flow velocity areas. A wide range of aerobic microorganisms act as a binder for the settled suspended solids. As the sludge accumulates, the aerobic microflora continues to produce substantial quantities of biomass within the deposited inorganic materials. The total mass of the deposit produces a barrier that retards the diffusion of oxygen through the mass. The aerobic microflora also consumes oxygen existing at the interface of the metal and the deposit. A substantial oxygen concentration gradient between the bulk water and the metal surface is developed. This readily provides ideal conditions for the initiation of under-deposit corrosion and, in many cases, the establishment of an environment in which anaerobic microorganisms, such as the sulfate-reducing bacteria (SRB) *Desulfovibrio sp.*, will grow. These anaerobic microorganisms characteristically initiate aggressive pitting corrosion sites and produce significant quantities of H_2S.

In addition to the pitting corrosion, this specific MIC problem produces undesired concentrations of soluble iron, iron sulfide, and hydrogen sulfide in the raw water.

Treated Process Water

Make-up water systems using pretreated water for showers, stock washing, consistency control, and stock dilution in both the pulp mill and the paper machine are potential sites for MIC.

Although extensive pretreatment may be done to remove impurities from process make-up water supplies, treatment to eliminate the inoculum source for microorganisms involved with MIC is often overlooked. Most of the suspended solids, and in some cases much of the dissolved solids, are removed by pretreatment. Filters, clarifiers, and water storage tanks used in water treatment plants provide ideal environments for microorganisms, such as *Sphaerotilus sp.* and *Siderocapsa sp.*, to grow. The population established in those environments contributes organisms capable of initiating MIC in other areas of the system in which the water is used. The amount of corrosion occurring in the make-up water system because of MIC may not be significant. However, because the treated process water is used at several widespread locations in the plant, the potential distribution of inoculum for MIC must be considered significant.

Economizer System

Both indirect heat exchanger economizers and direct warm water economizer systems using condensate for make-up are potential sites for MIC. Indirect heat exchanger economizers are often tube and shell heat exchangers with the process water on the shell side. The low-flow velocity or intermittent flow through the shell side of the heat exchanger provides an environment in which several types of microorganisms related to the MIC problem can grow. Often, a mixed microflora of slime-forming bacteria are associated with the MIC microflora. A low dissolved oxygen level in the warm water (30 to 40°C) can be depleted readily by the aerobic slime formers and can create an aerobic environment that supports the growth of sulfide-producing bacteria. The nonferrous materials of construction for the economizer system are not exempt from attack by MIC.

Direct warm-water economizer systems using condensate/treated process water for make-up are optimum sites for the growth of those MIC microorganisms that produce acids or concentrate sulfides and chlorides as metabolic byproducts. The iron and other metallic ions produced from contaminated condensate, as well as the material of construction of the economizer itself, contribute to the essential growth elements needed by the microorganisms. As a result of their growth, troublesome deposits are formed on equipment surfaces, suspended solids are dispersed in the bulk water, and anodic cells initiated by the microorganisms continue the corrosion process electrochemically. The latter is particularly significant when the metallurgy involved is stainless steel (SS) or other supposedly corrosion-resistant materials.

White Water (Back Water) Systems

Recycled white water constitutes the single greatest commodity used in pulping and papermaking processes. With few exceptions, the entire white water system is an environment that supports the growth of microorganisms that can produce MIC.

More than 85 percent of the water materials balance of an integrated pulp and paper mill is recycled water called "white water." As a result of recycling, the water contains high concentrations of dissolved solids, both organic and inorganic compounds, that supply the nutrients for a widely mixed microflora. Unclarified white water contains not only the nutrients but high quantities of fines, filler, and other suspended solids that provide an environment ideally suited for the uncontrolled growth of MIC microorganisms. MIC can occur specifically in low-flow velocity sites in the white-water system. Although the white water cycled in pulping and papermaking systems usually contains adequate oxygen to support the growth of most aerobic microorganisms, the growth of anaerobic bacteria, especially SRB, is prevalent under sludge caused by the settling of suspended solids and in "dead-flow" areas in the system. White water pH and temperatures are usually not limiting factors to the growth of the MIC microflora. The result of the growth of this microflora includes:

- Pitting or general corrosion on inside surfaces of suction boxes and foils,
- Pitting corrosion on back surfaces of screens and flow boxes,
- General corrosion in cylinder vats and vacuum disk/cylinder savealls,
- Pitting and general corrosion in pulp mill and paper machine consistency regulators,
- Tuberculation and general corrosion in paper machine white water trays,
- Pitting and tuberculation at random locations in pulp mill and paper machine white-water lines,
- Pitting and general corrosion in white-water retention tanks of flotation saveall systems,
- Pitting corrosion in stock deaerating equipment and deculators, and
- General corrosion, tuberculation, and occasional pitting in pulp mill decker filtrate tanks.

In addition to the deterioration of the equipment listed above, MIC often has a direct effect on the operating efficiency of the equipment. The development of odors and discoloration of stock and water in the white-water systems as a result of the growth of MIC microflora is a major concern.

Paper Machine Headboxes

Even a minimum of corrosion in headboxes cannot be tolerated. For that reason, most headboxes are typically constructed of materials such as SS that provide resistance to electrochemical corrosion. However, experience has shown that SS construction does not ensure resistance to MIC.

Poor sheet formation, stock lumps and stringers, dirt/rust specks in the sheet, and sheet discoloration are often the result of MIC inside the headbox. Pitting of SS and tuberculation of carbon steel components have been attributed to the growth of SRB and iron-oxidizing bacteria. Rough corroded surfaces not only provide a site for microbiological deposits to form in the headbox, they also provide an anaerobic environment for SRB to exist. MIC in headboxes is frequently associated with the uncontrolled growth of slime-forming bacteria and/or fungi that form dense mycelial growth in the low-flow areas. The deposit-forming microorganisms provide the environment that supports the growth of microorganisms directly involved with the corrosion process. As the corrosion progresses, the slime deposits are more difficult to control and, as the slime deposits become more extensive, the MIC becomes more difficult to control.

Paper Machine Suction Rolls

Suction roll and couch roll failures have increased significantly in recent years. Materials of construction of the rolls include SS, admiralty brass, and bronze. Based on knowledge of the metallurgy of these alloys, suction roll failure cannot be attributed to electrochemical corrosion/ fatigue mechanisms alone. Investigations have shown that MIC was involved in many of the failures.

As a result of recent investigations, at least two possible mechanisms for MIC failure of suction rolls have been proposed. One proposal relates to the uncontrolled growth of slime-forming bacteria, such as *Siderocapsa sp.*, that can initiate anodic corrosion sites on many metals, including brass and copper alloys. These bacteria produce an extracellular slime film that adheres to the surfaces of the suction roll shell, particularly on the concave surfaces and drilled suction holes. The biofilm is resistant to the shear and frictional forces encountered at this location and literally binds the bacterial growth to the metal surface. Few nonslime-forming microorganisms are able to survive in this environment, providing little competition to the growth of the slime formers. As the slime formers continue to grow, the surface of the metal is exposed to attack by exoenzymes and metabolic materials produced by the bacteria. Micropitting results and corrosion fatigue develops to the extent that the suction rolls fail.

A second proposed mechanism involves the formation of fiber plugs in the drilled suction holes. The plug blocks the flow of water and air through the drilled hole. This produces an anaerobic site at which SRB growth occurs and H_2S is produced. The sulfide interacts with the stressed metal surfaces of the drilled holes and alloy failure (including failures of SS and copper/brass) occurs and ultimately causes the suction roll to fail.

Cooling Water Systems

Problems associated with MIC in pulp and paper mill cooling water systems are similar to those encountered in the cooling water systems of all major industrial processes. Aerobic and anaerobic corrosion mechanisms described elsewhere occur in the steam condenser cooling system, in the turbine cooling system, and in the general process cooling system. Most modern pulp and paper mills are equipped with massive air conditioning systems that are also subject to MIC.

Fire Safety Water Systems

Pulp and paper mills store, handle, and process large quantities of highly combustible materials, i.e. paper, wood chips, waste paper, solvents, process chemicals, oxidizing bleaching chemicals, etc. Therefore, the plants must have extensive fire safety water systems. Most of the systems are wet sprinkler types that are filled with water at all times, and are stand-by closed loop water systems. For safety reasons, it is important that these systems be operational at all times. The environment within the fire safety water system provides optimum conditions for many of the iron-oxidizing bacteria and, in some cases, allows sulfide-producing bacteria to grow without limitation. The result of this growth may be actual pipe wall penetration by pitting corrosion. Often, iron sulfide or iron oxide deposits plug valves and sprinkler nozzles, making the system nonfunctional. Unfortunately, these situations can occur without warning and can result in the failure of the fire safety water system when it is needed most.

Prevention of MIC

Preventing the establishment of MIC in pulp and paper mill systems requires the coordination of many operational groups within the plant. It is usually more effective to take steps to prevent MIC than it is to mitigate MIC after it has been established. A coordinated plan involving the technical, engineering, and operational groups must be implemented before a significant degree of corrosion has developed. When the conditions under which MIC can occur are understood by each group, steps can be taken to eliminate those conditions.

The prevention plan should also include a sound program for deposits control, particularly focusing on the growth of microorganisms in general. Good housekeeping procedures are important steps to preventing the establishment of MIC because dirty systems will inevitably be troubled by MIC. Locating and isolating all the potential sites of inoculum source is necessary to chemically treat the system effectively. Possible sites include make-up water sources, recycled white water sources, contaminated stock systems, chemical additive systems, and "dead-end flow" sites in the system. The proper application of broad spectrum microbiocides will effectively keep the inoculum below that needed to establish MIC. Under these conditions, the microbiocides used for controlling conventional microbiological problems are effective in preventing MIC.

A systematic procedure for monitoring the presence of MIC inoculum and the establishment of corrosion sites should be established. Acidic chemical cleaning or "boil out" should be avoided. The use of a nonionic penetrant biodispersant as part of the routine microbiological control program should be considered. Special microbiocide applications should be made to the stock, broke, and white water systems when extended shutdowns occur. Treatment of the raw water system prior to pretreatment should not be overlooked. Technical review sessions discussing the technology of MIC corrosion control should be scheduled periodically with the appropriate operator and technical staff.

Mitigating an Existing MIC Problem

In the past, it was not uncommon to live with a MIC problem until it was necessary to replace or remove the equipment involved. It was believed that not much could be done to mitigate MIC. However, there has been significant progress toward eliminating MIC, or at least preventing the situation from getting worse. This process has been a result of several factors including:

- Development of procedures to more clearly identify MIC,
- Definition of the environmental conditions under which MIC occurs,
- Engineering advancements that reduce the potential for MIC,
- Development of barrier film corrosion inhibitors, compatible with papermaking processes, that can passivate MIC sites.
- Development of nonaggressive cleaning compounds and application procedures to remove deposits, and fouling materials associated with MIC,
- Engineering advancements that enable mitigation procedures to be performed,
- Development of biodispersants that enable microbiocide treatment to be effective at the sites at which MIC microflora exist, and
- Development of microbiocides that control the growth or eradicate MIC microflora without upsetting the biological equilibrium.

Specific procedures for mitigating MIC must be developed for each situation. It should also be recognized that mitigation cannot be accomplished in a short time period. However, in spite of the fact that each situation may be different, there are some basic principles that apply to all mitigation situations.

1. The system should be mechanically cleaned to remove all deposits and debris that can be discharged physically from the system, either by a systemwide cleaning or a localized cleaning where the MIC exists. Although the cleaning should be as thorough as possible, it should not be done in a manner that is mechanically aggressive to the system equipment. In some situations, a mild caustic/dispersant cleaning solution will facilitate this cleaning.

2. Following cleaning, flushing, and draining, all necessary maintenance and repair should be completed.

3. The system should then be filled with water to which a passivating nonionic penetrant dispersant is added.

4. After the system surfaces have been coated with the penetrant, a nonoxidizing microbiocide with specific efficacy against the MIC microflora should be applied to the water in the system. If possible, the water containing the penetrant and the microbiocide should be circulated for at least several hours. The specific time circulation required for maximum effectiveness of the mitiga-

tion procedure will depend on the specific situation.

5. *Never use aggressive chemical or mechanical cleaning procedures. The cleaning procedure must be performed passively to prevent activation of passivated corrosion sites and to prevent substituting a MIC problem for an electrochemical problem.*

Each step in the mitigation procedure must include provisions to passivate the corrosion process rather than to accelerate it. Unfortunately, many conventional corrosion inhibitors used for controlling corrosion in process water systems may not be compatible with the paper or pulp mill processes/materials of construction. Therefore, special consideration must be given to both the selection of treatment chemicals and the operating procedures used.

6. As indicated earlier, the mitigation procedure may involve repeating some steps or carrying the procedure "on-line" over an extended period of time. During the mitigation procedure, monitoring should be conducted to verify the elimination of the MIC microflora, and that passivation exists, and to confirm that no new MIC has been established.

7. Once the mitigation procedure has been completed, a well-coordinated program for the prevention of MIC should be initiated.

(Figures 4.1 - 4.11 appear in Color Plates - Section I, pages 85-87.)

Bibliography

1. Bonner, E.J. "What Causes Corrosion, and How It Can Be Controlled," *Pulp and Paper International* 11, 5(1969): pp. 55-59.
2. Bowers, D.F. "Corrosion Considerations for Alkaline Paper Making," *TAPPI* 71, 1(1986): pp. 62-66.
3. DeBerry, D.W., P.F. Ellis, "Inhibition of Localized Corrosion in Paper Mill White Water," CORROSION/88, paper no. 346, Houston, TX: NACE, 1988.
4. Grimm, D.T., C.G. Hollis, S.T. Threlkeld, "The Nature of Sulfate Reducing Bacteria Isolated from a High Temperature Kraft Paper Machine," *TAPPI Proceedings-1984 R and D Conference*, pp. 111-114, 1984.
5. Laliberti, L.H. "Corrosion of Stainless Steel in a Bleach Plant," *Proc. Conf. on Pulp and Paper Corrosion, 1974.* Houston, TX: NACE, 1974.
6. LaPonte, M., et al. "Corrosion of Stainless Steel in a Fine Paper Machine," *Pulp and Paper Canada* 80, 8(1979): pp. T251-T254.
7. Mueller, A., J.M. Muhonen, "Pitting Corrosion in Six Paper Machine Headboxes: Mechanisms and Prevention," *TAPPI* 55, 4(1972): pp. 589-598.
8. Stitt, J.B. "Monitoring and Controlling Wet-end Roll Corrosion in Paper Machines," *Pulp and Paper* (1977): pp. 22-25.
9. Thorpe, P.H. "Microbiological Corrosion of Stainless Steel in Paper Machines," *APPITA* 40, 2(1985): pp. 108-111.
10. Thorpe, P.H. "Corrosion Control in Paper Machines," *APPITA* 35, 6(1982): pp. 505-511.
11. Vestola, J., J. Korhonen, "Corrosion and Microbiological Action in Suction Rolls," *TAPPI* 59, 4 (1976): pp. 130-133.
12. Wolfson, L.L., R.J. Michalski, "The Incidence and Effects of the Anaerobic Bacteria Clostridium in Paper Mill Systems," *TAPPI* 47, 7(1964): pp. 197-199.

CHAPTER 5

MIC in Metalworking Processes and Hydraulic Systems

By H.W. Rossmoore and L.A. Rossmoore

Metalworking and Hydraulic Fluids

Introduction

Microbiologically influenced corrosion (MIC) in metalworking processes is well-documented.[1-4] Two factors contribute to the vulnerability of metalworking surfaces and hydraulic systems to corrosion, especially as influenced by microbial intervention. These are: the exposure of nascent metal as a result of the metalworking process or from piston wear (Table 5.1), and the introduction of water into both metalworking and hydraulic fluids. The former increases the tendency for metal oxidation while the latter guarantees an environment suitable for microbial growth. Fluids are ultimately designed with additives to accommodate for corrosion potential.

Metalworking Fluids

A number of classification schemes have been suggested for metalworking fluid (MWF), including one from the International Standards Organization (ISO)[5] and the Independent Lubricant Manufacturers Association (ILMA), as well as ASTM (Table 5.2), which are based on chemistry, function, application, or a combination of all three. Although so-called straight oils continue to be used for certain low-speed, high-friction operations, practically all production metalworking processes use water-based fluids.

Hydraulic Fluids

Hydraulic fluids (HF) must provide anti-wear as well as corrosion protection in the systems in which they are employed. The introduction of water-based HF has two purposes: to reduce flammability (e.g., in mine shafts) and to reduce the level of petroleum-based products in the fluid.

There are two major types of water-containing HF, both of which can have microbiological problems. The first was developed to have the same viscosity as traditional hydraulic oil to satisfy design criteria of vane pumps; they are chemically water-in-oil emulsions (so-called invert emulsions), contain 40% water and 60% oil, are nonflam-

TABLE 5.1
Selected Metalworking Operations Involving Fluids

I. REDUCTION IN NET SHAPE
Broaching, Tapping, Threading, Gear Cutting Fluid Recirculated
Drilling, Milling, Sawing, Planing
Grinding
II. CHANGE IN SHAPE—DEFORMATION
Sheet Rolling (Hot and Cold),
Can Drawing and Ironing Fluid Recirculated
Stamping Once Through
Tube Drawing
III. METALWORKING FLUID FUNCTION—DEPENDENT ON I AND II
Reduce Heat—Cool Tool and Workpiece
Increase Lubricity
Carry Off Chips, Fines, and Swarf
Inhibit Corrosion of Tool and Workpiece

TABLE 5.2

Chemical and Functional Classification of Metalworking Fluids (ASTM D2881-73)
(Reprinted with permission from ASTM)

I. Emulsions

A. WATER IN OIL (W/O) (Invert Emulsions)—Used as hydraulic fluids and in operations involving metal deformation; contain 60 to 70% mineral oil; proprietary emulsifiers, H_2O; anti-wear additives for hydraulics; avg. pH 8 to 8.5

B. OIL IN WATER (O/W) (Soluble Oils)—Used universally for almost all types of metalworking; contain 10 to 85% mineral oil, 10 to 50% extra pressure (EP) additives (e.g., chlorinated or sulfurized paraffins), 10 to 25% emulsifiers; fatty acids, esters, biocides, corrosion inhibitors occasionally; avg. pH 8 to 8.5.

II. Chemical Solutions (Synthetics)

A. TRUE SOLUTIONS (Synthetics = No Oil)—Alkanolamines 20%, borates 4 to 20%, biocides, miscellaneous corrosion inhibitors, H_2O 50 to 70%; avg. pH 9 to 9.5.

B. COLLOIDAL SOLUTIONS (Microemulsions)—Fatty acids mechanically dispersed with non-ionic surfactants 15 to 20%, corrosion inhibitors 15 to 30%, biocides, H_2O 50 to 70%; avg. pH 9 to 9.5.

III. Preformed Emulsions

A. (Semi-Synthetics)—Mineral oil 10 to 40%, emulsifiers 5 to 10%, EP additives 10 to 40%, H_2O 20 to 50%, corrosion inhibitors 10 to 20%, biocides, avg. pH 9 to 9.5.

mable, and have the same viscosity as oil. The newer group of products is referred to as high water-based fluids (HWBF), high water-content fluids (HWCF), or 95 to 5 fluids, indicating the most common water:lubricant ratio. These are primarily limited to piston pumps since their viscosity is too low for vane pumps designed for oil. Both the invert emulsions and 95 to 5 fluids are very susceptible to microbial growth.[6,7] As a result, a standard has been developed for evaluating biocides in those fluids.[8]

TABLE 5.3

Types of Corrosion Inhibitors Used in Metalworking Fluids and Hydraulic Fluids

I. Passive Inhibitors

A. OXIDIZING TYPE
Nitrite, Chromate

B. NONOXIDIZING TYPE
Borates, Molybdate

II. Polar Organic (Adsorption Type)

Amines (Amine Borates, Primary, Secondary, and Tertiary Alkanolamines)
Amides, Fatty Acids, Sulfonates, Miscellaneous Carboxylates

III. Barrier Layer

Oils as Oleophiles in O/W Emulsions

Cause and Mitigation of MIC

The protection of exposed metal surfaces from corrosion is a requirement of both MWF and HF, and the appropriate types of corrosion inhibitors are listed in Table 5.3. In soluble oils (Table 5.2), the primary corrosion protection is afforded by a barrier layer (Table 5.3). Biodegradation of these fluids begins with hydrocarbon oxidation step-wise to fatty acids, resulting in lowered pH. Concomitantly, there may be utilization of the emulsifier and with either or both events, low pH or loss of emulsifier, the emulsion splits into two phases (Figure 5.1) with subsequent loss of barrier film protection; contact with air-saturated water then results in corrosion. (Figures 5.1-5.8[b] appear on pages 87-89.)

Most MWFs (Table 5.2) contain chemical corrosion inhibitors.[9-12] The simplest and perhaps the least expensive of these inhibitors, $NaNO_2$, has lost favor, especially in combination with amines, because of the potential for nitrosamine formation.[13] However, $NaNO_2$ may still be used in certain limited situations such as passivation of cold-rolled steel, even though it is relatively biodegradable.[14]

Microbial Populations

Mechanism

Microbial activity can influence corrosion in metalworking and hydraulic systems, both directly and indirectly as described for oil/water barrier film degradation. Mechanisms include direct anaerobic oxidation, differential oxygen concentration effects, and interference with action of corrosion inhibitors.

Populations

The organisms contributing to MIC include aerobic bacteria, anaerobic sulfate reducers, yeasts, and filamentous fungi. It is the aerobic populations that are the pioneers in microbial succession. These aerobes are primarily *Pseudomonads,* [15] and, although not all species are hydrocarbon oxidizers, together they complete the sequence of breakdown from hydrocarbon to acid.[16,17] These organisms also pave the way for sulfate reducers[18-20] and fungi.[21,22,15]

Between production shifts, when MWFs are not flowing or being aerated, a natural consequence is the lowering of the oxidation potential by the respiration of the aerobes sufficient for resident sulfate reducers to grow (Figure 5.2). Anaerobic corrosion can then occur at the bottom of steel reservoir sumps, under tool ways, and on tool and workpiece surfaces.

In an extensive study of MWFs from more than 100 systems, dependence of both fungi and sulfate reducers on prior growth of anaerobic bacteria was noted.[15] These former groups are capable of survival and frequently grow after the death of the aerobic population.[20,22-24]

Attached Microbes in Corrosion

Generally, the most acceptable, universal microbial activity related to corrosion is the phenomenon of microbial adhesion or biofilm formation.[25] In the metalworking areas, with MWF, biofilm formation is easily demonstrable with both bacteria and fungi. Cast iron chips in a contrived MWF biodeterioration test accumulate an attached bacterial population, primarily *Pseudomonads*, although sulfate-reducing bacteria (Table 5.4) are found accompanying the aerobic population. The attached populations are classified as loosely or tightly bound. The former are removed by gentle shaking, and the latter by sonic oscillation. Both were compared with the levels of free-floating (planktonic) bacterial populations. Notice that the planktonic levels drop as the attached levels rise.

TABLE 5.4

Relationship Between Sessile and Planktonic Microbial Population on Metal/Metalworking Fluid Environment [(1)]

Time (Days)		Colony-Forming Units: Aerobic Bacteria	Colony-Forming Units: SRB
	Planktonic initial/mL	1.4×10^7	$>10^5$
	Planktonic final/mL	2.0×10^5	$>10^5$
3	Loose/g	10^5	10^4
	Bound/g	1.3×10^8	$>10^5$
5	Loose/g	2.7×10^6	$>10^5$
	Bound/g	3.2×10^8	$>10^5$
7	Loosely bound/g	8.5×10^4	$>10^5$
	Tightly bound/g	1.4×10^9	$>10^5$

(1) Contaminated mixed metalworking fluids were trickled through a mixture of chips, fines, and swarf and recirculated for one week. Microbial counts were taken of fluid at time zero and at one week (planktonic). At times listed, one g sample of metal mixture was coarse filtered and suspended in 100 mL of sterile distilled water and shaken vigorously for 30 s (loosely bound) or subjected to 40,000 cycles/s oscillation for 5 minutes (tightly bound). These latter samples are considered sessile populations.

Relationship of Biocides and Corrosion Inhibitors

The control of corrosion by control of microbial activity is further, but indirect, evidence of MIC in MWFs and HF. The selection of biocide must be based on compounds registered specifically for this use.[4] Biocides for MWF have been placed in the high exposure category by the Environmental

Protection Agency (EPA), and the subsequent reregistration of a number of these are in doubt. Because the need in these fluids for both biocides and corrosion inhibitors has been demonstrated, the possibility of their interaction is an ongoing concern. Obviously, antagonism between them is to be avoided and, therefore, must be evaluated in corrosion-inhibition tests.

Interaction of Corrosion Inhibitors and Biocides

Interaction depends considerably on the mode of action of both compounds. A number of MWF biocides, including glutaraldehyde[26,27] and methylchloroisothiazolones (MCI),[28,29] are partially neutralized by amine corrosion inhibitors. Although reaction with biocides would reduce the level of corrosion inhibition, the ratio of 10 to 20 of amine corrosion inhibitor to 1 of biocide would not significantly affect corrosion protection.

There is also information on the synergistic biocidal activity produced by known corrosion inhibitors. In MWFs (Table 5.5), Na_2MoO_4 increases the efficacy of MCI but only against sulfate-reducing bacteria (SRB). However, Na_2MoO_4 appears to reduce the activity of dibromonitrilopropionamide against aerobic bacteria (predominantly *Pseudomonads*). This latter biocide has been prominent in preserving fluids used in the forming of tin-plated steel and aluminum beverage cans, an operation where microbial and corrosion control standards are the metalworking industry's highest.

TABLE 5.5

Effect of Sodium Molybdate on Metalworking Fluid Biocide Activity Against *Desulfovibrio desulfuricans.*[(1)]

	Na Molybdate (ppm)	MCI[(2)]	SRB (cfu/mL)
Control	—	—	10^6
Test #1	500	—	10^6
Test #2	1,500	—	10^5
Test #3	—	5	10^5
Test #4	500	5	10^5
Test #5	1,000	5	10^2
Test #6	—	10	10^5
Test #7	500	10	10^3
Test #8	1,000	10	10^2

[(1)] *Desulfovibrio desulfuricans* mid-continent strain was grown in API broth and acclimated to a 5% soluble oil emulsion. Estimation of *D. desulfuricans* (SRB) was with a semi-quantitative proprietary test kit.[30,31]

[(2)] The biocide concentrate contained 14.5% active, 10.98% 5-chloro-2-methyl-3(2H)-isothiazalone and 3.6% 2-methyl-3(2H)-isothiazalone.

Biocidal Activity of Corrosion Inhibitors

A number of compounds, primarily used as corrosion inhibitors, have some selected biocidal activity. Perhaps the best example are the borates[32] and amine derivatives of borates.[33] They inhibit bacteria at recommended corrosion use levels effectively but not fungi to the same extent. This differential activity creates the false impression initially of a bioresistant fluid.

In a bioresistance study on a supposedly biocide-free MWF, bacterial populations were controlled.[34] Subsequently, it was learned that this fluid contained mercaptobenzothiazole (MBT) as a copper antioxidant.

In a separate study, MBT reduced the activity of lactic dehydrogenase from sulfate-reducing bacteria, suggesting a molecular basis for its dual activity.[35] In addition, hydraulic oil fungal growth was not present in a control oil containing 0.2% morpholine that was incorporated solely as a corrosion inhibitor (Figure 5.3).

Bennett and colleagues investigated an extensive list of putative corrosion inhibitors based on alkanolamines as potential MWF biocides.[36-38] In one study, Bennett found 2(Namyl) ethanolamine was effective in controlling both bacteria and fungi in 10 of 13 MWFs.[37] However, no parallel corrosion inhibition tests were conducted.

Corrosion Protection by Biocides

Some biocides also contribute to corrosion control by helping to maintain alkaline reserve. A number of amine-formaldehyde condensate products, including 1,3,5-hexahydro tris (2 hydroxyethyl) triazine, have been formulated with MWF concentrates taking this property into consideration. One derivative, 4,4 dimethyloxazolidine, was originally used for corrosion protection; its subsequent EPA registration as a MWF biocide recognized that latter activity.

Corrosivity of Biocides

Many biocides are potentially corrosive, especially oxidizing compounds and many containing halogens, thus *a priori* limiting their use in MWFs, although there are EPA-registered products for MWF that chemically fit that description. However, others in certain conditions (Tables 5.7 and 5.9) are demonstrably corrosive.

MIC Environmental Factors

Metalworking - Effect of Hard Water

In a study to relate microbial activity, chemical breakdown, and engineering function, two factors contributed to increased microbial growth.[39] These were the increase in water hardness (divalent cation) and the use of reclaimed oil stocks for formulation of the soluble oil. Those variables perceptibly increased sulfate reducer levels and resulted in failure in a modification of the ASTM D4627-86 cast iron chip corrosion test.[40] This modification involved decanting MWF after a five-minute contact and allowing the wet chips to be exposed to air overnight.

In the actual test, the chips remain submerged overnight and then the MWF is decanted. The results (Figure 5.4) clearly showed that the modification is a severe challenge and that the decanted systems offered no protection against corrosion.[39]

Control fluids without biodegradation offered corrosion protection, whether submerged overnight or exposed to air, after five minutes of fluid contact. It was apparent that microbial activity was responsible, in some measure, for the loss of corrosion protection.

Effect of Diluent Contamination on Stability of Corrosion Inhibitors

Microbiological involvement in black pitting on cold-rolled steel was investigated. Pitted surface samplings were negative, either for sulfate reducers or for large or unusual numbers of aerobic bacteria or fungi. One clue involved $NaNO_2$ solution into which the fresh rolled surface was dipped. The concentrated commercial $NaNO_2$ solution had been diluted with untreated Lake Michigan water for use, and when associated with the pitted steel had "soured" (i.e., pH on the acid side) with perceptibly higher bacterial counts than normal $NaNO_2$ solution, and with 99 percent reduction in $NaNO_2$ (Table 5.6).

Two studies were performed to verify MIC. In the first study, various solutions containing sulfate reducers, isolated from contaminated $NaNO_2$, were sandwiched between two steel coupons. The solution was surrounded by vaspar (mineral oil and petroleum jelly) to inhibit oxygen diffusion, and the sandwich was clamped shut. The results are shown in Figure 5.5.

A parallel experiment carried out at the company metallurgy laboratory, without the sulfate reducer culture but with all the permutations, resulted in rust-corrosion but without any black stain.

Despite the apparent positive nature of the

TABLE 5.6

Source and Level of Microorganisms Associated with Soured $NaNO_2$ Solution

	pH	cfu/mL[(1)] Aerobic Bacteria	cfu/mL[(1)] SRB	ppm NO_2
Lake Michigan Diluent	8.4	3×10^3	$\geq 10^4$	
Fresh NO_2 Solution	8.8	4×10^3	<10	10,000
Sour NO_2	6.6	1.2×10^6	10^5	<100

[(1)] Aerobic bacteria quantified by standard plate count method; SRB by a semi-quantitative proprietary test kit.[30,31]

sandwich experiment, direct proof of MIC is lacking in this case. Scanning field-stained steel with Auger spectroscopy revealed no levels of P that could be related to biological deposition. However, the use of $NaNO_2$ solutions, subsequently treated with biocide, did result in the control of the black pitting problem in the field.

Effect of MWF Chemistry and Microbial Levels on Corrosion

A large automotive manufacturing division requires MWF vendors to qualify their products in an 8- to 10-week lab performance test to relate corrosion, microbial growth, and fluid chemistry. One protocol is a modification of ASTM D3946-86 ("Standard Method for Evaluating the Bioresistance of Water-Soluble Metal-Working Fluids")[41] supplemented by the ASTM D4627-86 cast iron chip corrosion test.[40] The results in Table 5.7 suggest that corrosion, when positively correlated with microbial levels, takes several weeks to show up. Even with this select sample of fluids there is no perfect agreement between high counts and corrosion. The stability of the corrosion inhibitor package, the biodegradation products of the MWF, and the microbial species contribute to the final relationship.

TABLE 5.7

Relationship Between Microbial Levels, Corrosion, and Metalworking Fluid Stability (Soluble Oil)

Exposure		**Fluid A**		**Fluid B**	
(Weeks)		**With Biocide**	**Without Biocide**	**With Biocide**	**Without Biocide**
1	pH	9.0	8.9	8.7	8.6
	AB(1) cfu/mL	$<10^3$	$<10^3$	$<10^2$	3×10^6
	Fungi(2) cfu/mL	<1	3×10^2	<1	2.4×10^3
	SRB(3) est./mL	$\geq 10^3$	$\geq 10^3$	$\geq 10^3$	<1
	Corrosion(4)	2	2	2	1
3	pH	9.0	8.7	8.7	8.6
	AB cfu/mL	$<10^2$	$<10^2$	$<10^2$	2.5×10^5
	Fungi cfu/mL	$<10^2$	4.2×10^4	<1	10^3
	SRB est./mL	$>10^2$	$>10^2$	<1	<1
	Corrosion	2	1	1	1
5	pH	8.8	8.6	8.7	8.2
	AB cfu/mL	$<10^2$	10^6	$<10^2$	8×10^7
	Fungi cfu/mL	<1	2×10^3	10^2	2×10^3
	SRB est./mL	<1	<1	<1	<1
	Corrosion	2	3	2	2
7	pH	8.8	8.4	8.6	8.2
	AB cfu/mL	$<10^2$	2.5×10^7	$<10^2$	1.2×10^8
	Fungi cfu/mL	<1	2×10^3	<1	2×10^3
	SRB est./mL	<1	<1	<1	<1
	Corrosion	2	5	2	5

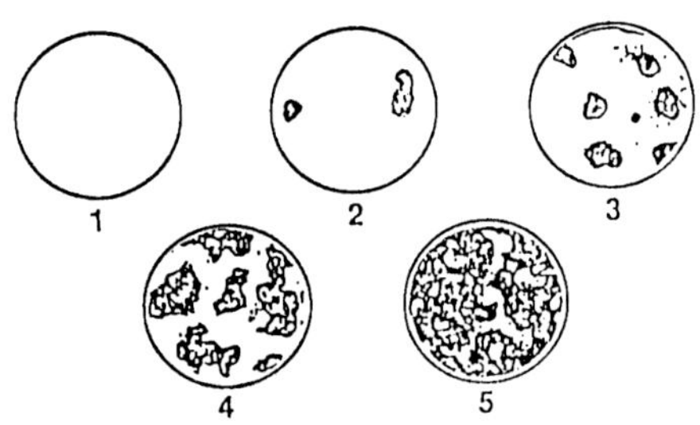

(1) AB = aerobic bacteria, quantified by standard plate count procedure.
(2) Fungi quantified by standard plate count procedure.
(3) SRB = sulfate-reducing bacteria, quantified by a semi-quantitative proprietary test kit.[30,31]
(4) Key to Corrosion Rating (Reprinted by permission of the Society of Tribologists and Lubrication Engineers[33]):
1 = no rust spotting of paper; 2 = slight spotting;
3 = moderate rusting; 4 = severe rusting; 5 = gross rusting

Hydraulic Fluids

Long-Term Incubation Impact on Hydraulic Oil Corrosivity

Hydraulic oils should offer minimal microbiological problems because of the absence of water. Given time, entrained water will be sufficient to permit growth. In a contrived study with uninoculated seawater overlaid with oil, a macroscopic biomass (Figure 5.4) appeared after two months of incubation. The dominant organism was identified as *Cladorsporium resinae*. As mentioned earlier, the presence of morpholine as a corrosion inhibitor prevented the growth. This biomass, in a hydraulic system attached to an internal surface, would induce corrosion via oxygen concentration differences.

Biocide Control of Filter Fouling Involving Invert Emulsion HF

A fouled filter (Figure 5.6) from a steel hydraulic press had a mixed microbiota of bacteria, yeast, and molds, and revealed surface corrosion (Figure 5.7) after removal of the biomass.[6] In an experimental study that incorporated biocide, control of microbial populations also controlled biofouling of filters and presumably with less chance for corrosion induction (Figures 5.8[a] and 5.8[b]).[42]

TABLE 5.8

Comparative Chemical Analysis of Mine Diluent Water With Detroit City Water[43]

Chemical	City of Detroit (ppm)	Mine 1 MCC (ppm)	Mine 2 Ranger (ppm)
Aluminum	0.11	0.15	0.46
Iron	0.015	0.10	0.07
Magnesium	7.0	12.0	9.5
Calcium	27.0	14.0	28.0
Sodium	4.6	300.0	74.3
Silica	1.4	ND[(1)]	9.0
Sulfate	22.0	175.0	70.0
Chloride	7.0	20.0	21.0
Phosphate	0.01	0.015	0.12
Nitrate	0.16	1.15	0.15
Nitrite	0.001	0.001	0.003
Carbonate	0.02	0.02	0.08
Bicarbonate	74.0	339.0	81.0
Fluoride	1.05	0.45	<0.2
Chemical Oxygen Demand	3.6	5.7	3.55
Dissolved Solids	125.0	884.0	301.0
pH	7.5	8.6	8.6

[(1)]ND = not determined

Effect of Water Quality and Poor Maintenance in an HWCF System

Fluid failure in a long wall mining operation is a striking example of the contribution of environmental factors, poor maintenance, and ignorance of microbiology to a corrosion problem with potentially disastrous results.[43] Long wall mines are excavated after exposing a horizontal coal seam. The horizontal shaft ceiling is supported by a series of hydraulic shield supports all fed by a single hydraulic sump. Reports of frozen pistons with attendant corrosion led to the study summarized here. An experimental system involving simulated filtration and continual fluid exposure to four metals (zinc, iron, copper, and brass) was set up with candidate HWCF. Fluids were inoculated with an adapted mixture of organisms, isolated from mine water, used to dilute the HWCF concentrate and selected biocides were used in parallel. Water quality was essentially poor when compared to city water (Table 5.8), and it appears to be a contributing factor in both corrosion and microbial growth. During a three-week period, pH, microbial levels, and appearance were monitored. Corrosion of metals was evaluated at the conclusion (Table 5.9). Again, it is shown that corrosion can have multifaceted causes.

It appears that location of mine (i.e., water), nature of fluid, and the biocide collectively contributed to corrosion. For example, Mine RA with Fluid I and no biocide appears to have the least corrosion, with reasonably high microbial levels. However, Mine MCC with Fluid II and Biocide B showed corrosion of three metals despite no detectable microbial levels. Mine MCC water is apparently incompatible with Biocide B; Biocide A in Fluid I with MCC prevents both growth and corrosion. Note that the pH at 9.0 may be a contributing factor.

TABLE 5.9

Effect of Biocide and Water Quality on Corrosivity of Contaminated Hydraulic Fluid[(1)] [43]

Fluid	Mine	Biocide	Iron	Copper	Brass	Zinc	pH	Aerobic Bacteria (cfu/mL)	Fungi (cfu/mL)	Sulfate-Reducing Bacteria (cfu/mL)
			Findings After Three Weeks							
I	MCC	None	+2[(4)]	+1[(3)]	+2	—[(2)]	8.5	10^7	3.5×10^3	$>10^5$
		A	—	—	+1	—	9.0	<1	<1	<10
		B	+3[(5)]	+3	+1	—	8.7	5×10^7	<1	<10
	RA	None	—	—	+1	—	8.0	4×10^6	3×10^2	$>10^4$
		A	—	+3	+3	—	8.6	<1	<1	<10
		B	+1	+1	+2	—	8.2	5×10^2	<1	$\geq 10^3$
II	MCC	None	+1	+1	+3	—	8.3	3×10^7	2×10^3	$>10^5$
		B	+3	+1	+2	—	8.7	<1	<1	<10
III	RA	None	—	+1	+2	—	7.6	2×10^7	2×10^4	$>10^4$
		A	—	+3	+3	—	8.4	<1	<1	<10
		B	+2	+2	+2	—	8.0	8×10^3	<1	$\geq 10^3$

(1) A protocol was carried out with weekly inoculation with all three types of organisms. Pieces of four metals representing those in affected mining equipment were placed in each quadrant of a four-section 100 mm petri dish and placed in the bottom of a 500 mL beaker containing 350 mL of test fluid diluted with mine water. Aeration was continual for the three-week test period.
(2) - = no visible corrosion.
(3) +1 = slight but detectable corrosion.
(4) +2 = some metal discoloration, more pronounced than +1.
(5) +3 = significant metal discoloration and slime coating indicating some metal deterioration.

Summary

Microbiological involvement in corrosion in the industries involving metalworking and hydraulic systems has been adequately reported. No other industrial area has a greater possibility for corrosion that exists in the large variety of metal types with freshly cut and exposed active surfaces after metalworking. The varied organic lubricants, wetting agents, and corrosion inhibitors are all relatively biodegradable, and, coupled with continual environmental contamination of these aqueous systems, the potential for microbial growth is maximal. This eventually leads to all of the classical causes of metallic corrosion.

References

1. E.O. Bennett, "The Biology of Metalworking Fluids," *Lubrication Engineering* 28(1972): pp. 237-247.
2. E.C. Hill, O.M. Gibbon, "Corrosiveness of Emulsions," *Metal Finishing* 15(1969): p. 395.
3. E.C. Hill, I.O. Penberth, "The Microbiology of Steel Mills and Its Relation to Sheet Corrosion," *Metals and Materials* 2(1968): pp. 359-366.
4. H.W. Rossmoore, "Microbial Degradation of Water-Based Metalworking Fluids," *Comprehensive Biotechnology* 3 (New York, NY: Pergamon Press, 1986), pp. 249-269.
5. ISO Standard 6743/7(E), "Lubricants, Industrial Oils, and Related Products (Class L)" Classification—Part 7: Family M (Metalworking) (Geneva, Switzerland: ISO, 1986).
6. H.W. Rossmoore, K. Szlatky, "Characterization of the Microbial Flora of Invert Emulsion Hydraulic Fluids," *International Biodeterioration Bulletin* 13, 4(1977): pp. 96-100.
7. L.A. Rossmoore, "Microbiology of High Water-Based Hydraulic Fluids," *Hydraulics and Pneumatics* 34, 9(1981): pp. 86-87.
8. ASTM Standard E979-91, "Standard Test Method for Evaluation Antimicrobial Agents as Preservatives for Invert Emulsion and Other Water Containing Hydraulic Fluids," *Annual Book of ASTM Standards* (Philadelphia, PA: ASTM, 1991).
9. H.W. Bernhardt, R. Helwerth, H. Hoffman, "The Application of Electrochemical Methods for the Evaluation of Corrosion Inhibitors in Aqueous Metalworking and Cooling Water Systems," *Lubrication Engineering* 35, 1(1979): pp. 36-41.

10. C.J. Fette, "Sodium (Nitrite) Replacement—The State of The Art," *Lubrication Engineering* 35, 11(1979): pp. 625-627.
11. M.S. Vukasovich, "Sodium Molybate Corrosion Inhibition of Synthetic Metalworking Fluids," *Lubrication Engineering* 36, 12(1980): pp. 708-712.
12. M.S. Vukasovich, "Rust Protection of Synthetic Metalworking Fluids with Nitrite Alternatives," *Lubrication Engineering* 40, 8(1984): pp. 456-462.
13. T.Y. Fan, J. Morrison, D.P. Roundbehler, D.H. Fine, R. Ross, W. Miles, N.P. Sen, "N-nitrosodiethanolamine in Synthetic Cutting Fluids: A Part-Per-Hundred Impurity," *Science* 196(1977): pp. 70-71.
14. D.G. Lundgren, A. Kriskszens, "Bacterial Destruction of Sodium Nitrite in Open Cooling Water Systems," *Applied Microbiology* 7(1959): pp. 292-300.
15. H.W. Rossmoore, L.A. Rossmoore, C.E. Young, "Microbial Ecology of an Automotive Engine Plant," in *Biodeterioration 5, Proceedings of the Fifth International Biodeterioration Symposium* (New York, NY: Plenum Publication Corp., 1987), pp. 255-268.
16. H. Pivnick, M. Fuller, H. Graham, S. Uyeno, "Biological Oxidation of Soluble Oil Emulsions," *Lubrication Engineering* 11, 2(1955): p. 96.
17. L.R. Sabina, H. Pivnick, "Oxidation of Soluble Oil Emulsions and Emulsifier by *Pseudomonas oleovorans* and *Pseudomonas formicans*," *Applied Microbiology* 4(1956): pp. 171-175.
18. G.J. Guynes, E.O. Bennett, "Bacterial Deterioration of Emulsion Oils. I. Relationship between Aerobes and Sulfate-Reducing Bacteria in Deterioration," *Applied Microbiology* 7(1959): pp. 117-121.
19. D.L. Isenberg, E.O. Bennett, "Bacterial Deterioration of Emulsion Oils. II. Nature of Relationship between Aerobes and Sulfate Reducing Bacteria," *Applied Microbiology* 7(1959): pp. 121-125.
20. H.W. Rossmoore, "Correlation of Coliform and Anaerobic Sulfate Reduction with Deterioration of Cutting Fluids," *Lubrication Engineering* 18, 5(1962): pp. 226-229.
21. H.W. Rossmoore, G.H. Holtzman, L.A. Rossmoore, "Microbial Succession in Cutting Fluid Milieu," *Abstracts of the Annual Meeting-1972* (Washington, DC: American Society for Microbiology, 1972), p. 3.
22. H.W. Rossmoore, G.H. Holtzman, "Growth of Fungi in Cutting Fluids," *Developments in Industrial Microbiology* 15(1974): pp. 273-280.
23. H.W. Rossmoore, G.H. Holtzman, L. Kondek, "Microbial Ecology With A Cutting Edge," *Proceedings of the Third International Biodegradation Symposium* (London, UK: Applied Science Publishers, 1976), pp. 221-232.
24. K.W. Vedder, H.W. Rossmoore, K.W. Rossmoore, "The Detection of Airborne Sulfate-Reducing Bacteria from Metalworking Fluids," in *Biodeterioration 6, Proceedings of the Sixth International Biodeterioration Symposium* (UK: C.A.B. International), pp. 453-459.
25. J.W. Costerton, K.J. Cheng, G.G. Geesey, T.I. Ladd, J.C. Nickel, M. Dasgupta, T. S. Marrie, "Bacterial Biofilms in Nature and Disease," *Annual Review of Microbiology* 41(1987): pp. 435-464.
26. J. Leder, "Glutaraldehyde: A New Microbicide for Metalworking," *Lubrication Engineering* 43, 8(1986): pp. 666-671.
27. K. Rossmoore, "The Microbial Activity of Glutaraldehyde in Chain Conveyor Lubricants," in *Biodeterioration 7, Proceedings of the Seventh International Biodeterioration Symposium* (Cambridge, UK: Elsevier Applied Science Publishers, 1988), pp. 242-247.
28. A.B. Law, E.S. Lashen, "Microbiocidal Efficacy of a Methylchloro-Methyl-Isothiazolone/Copper Preservative in Metalworking Fluids," *Lubrication Engineering* 47, 1(1991): pp. 25-31.
29. M. Sondossi, V. Riha, H.W. Rossmoore, "Factors Controlling the Biocidal Activities of Formaldehyde and Formaldehyde Condensates/Isothiazolone Mixtures," *Abstracts of the Society for Industrial Microbiology Annual Meeting* (Arlington, VA: Society for Industrial Microbiology, 1988), p. 47.
30. L.A. Rossmoore, J.W. Wireman, H.W. Rossmoore, "Rapid Field Method for the Detection and Enumeration of Sulfate Reducers," in *Biodeterioration 6, Proceedings of the Sixth International Biodeterioration Symposium* (UK: C.A.B. International, 1986), pp. 413-419.
31. R.E. Tatnall, K.M. Stanton, R.C. Ebersole, "Testing for the Presence of Sulfate-Reducing Bacteria," *Materials Performance* 27, 8(1988): pp. 71-80.
32. C.A. Sluhan, "Stable Aqueous Cutting Fluid," U.S. Patent 2,999,064 (1961).
33. P.M. Hernandez, I.C. Larsen, C.R. Petersen, "The Various Factors Involved in Developing a Microbial-Resistant Aqueous Cutting Fluid," *Lubrication Engineering* 41, 7(1985): pp. 425-429.
34. H.W. Rossmoore, P. Sceszny, L.A. Rossmoore, "Evaluation of Source of Bacterial Inoculum in Development of a Cutting Fluid Test Procedure," *Lubrication Engineering* 33, 7(1977): pp. 372-377.
35. M.H. Czechowski, H.W. Rossmoore, "Effect of

Selected Industrial Biocides on Lactate Metabolism in Desulfovibrio desulfuricans," *Developments in Industrial Microbiology* 22(1981): pp. 797-804.
36. E.O. Bennett, "The Antimicrobial Properties of Diethylene Triamines in Metalworking Fluids," *International Biodeterioration Bulletin* 14(1978): pp. 21-29.
37. E.O. Bennett, "Corrosion Inhibitors as Preservatives for Metalworking Fluids—Ethanolamines," *Lubrication Engineering* 35(1979): pp. 137-144.
38. E.O. Bennett, M.C. Adams, G. Tavana, "Antimicrobial Properties of Butanolamines and Propanolamines in Metalworking Fluids," *Journal of General Applied Microbiology* 25(1979): pp. 63-69.
39. G.H.M. Holtzman, H.W. Rossmoore, E. Holonik, M. Weintraub, "Interrelationships Between Biodeterioration, Chemical Breakdown, and Function in Soluble Oil Emulsions," *Developments of Industrial Microbiology* 23(1982): pp. 207-216.
40. ASTM Standard D4627-86, "Standard Test Method for Iron Chip Corrosion for Water Soluble Metalworking Fluids," *Annual Book of ASTM Standards* (Philadelphia, PA: ASTM, 1986).
41. ASTM Standard D3946-91, "Standard Method for Evaluating the Bioresistance of Metalworking Fluids," *Annual Book of ASTM Standards* (Philadelphia, PA: ASTM, 1991).
42. H.W. Rossmoore, "Heterocyclic Compounds as Industrial Biocides," *Developments in Industrial Microbiology* 20(1979): pp. 41-71.
43. L.A. Rossmoore, "Poor Quality Water Ruins HWBFs," *Coal Age* 89, 4(1984): p. 282.

MIC in the Oil Field

By G.L. Horacek, E.S. Littmann,[1] and R.E. Tatnall

Introduction

Of major industries, the petroleum industry probably suffers most from microbiologically influenced corrosion (MIC) and related biological problems. These MIC problems are not spectacular, nor do they raise the kind of safety and environmental concerns that make front page news stories. Yet the annual cost to field operations is immense.

Historically, preventing MIC problems has been a challenge within the industry. MIC problems are characteristically difficult to diagnose, treat, and monitor, especially "downhole." The industry has also been slow to recognize MIC's effect on operations.[1] Finally, petroleum-producing companies often lack the expertise for detecting and treating MIC problems. They routinely depend on chemical service companies to correctly handle their MIC problems. Unfortunately, these service companies have also been slow to recognize the impact of MIC on hydrocarbon production and to develop appropriate technology.

Wide disagreement exists among petroleum producers, microbiologists, and corrosion specialists about the magnitude of the industry's biological problems and the relative value of treating them. One extreme position states that most production problems, including reservoir souring, injection well plugging, sucker rod failures, and pitting of tubing, casing, aboveground tanks, and piping are biological problems. The problems are thus preventable by controlling microbial populations.

The opposite extreme is that these problems result from thermodynamic and physical/chemical phenomena within the production system. The problems may be preventable, but not by antimicrobial programs. The truth probably lies somewhere between these two extremes.

The following discussion outlines the state of MIC knowledge within the oil field. It covers such diverse issues as:

- Sources of microbial contamination,
- Common MIC symptoms,
- How MIC problems are currently handled,
- How biocides are applied, and
- How MIC and biocide applications are monitored.

It is concluded that there are a number of reasons the petroleum industry suffers from MIC. These reasons fall into three related categories:

- Industry operational practices,
- A failure to recognize MIC's impact, and
- The lack of concise and standard criteria for diagnosing, controlling and monitoring MIC.

Oilfield Contamination by Microorganisms

Microorganisms capable of causing MIC are commonly present in oilfield surface equipment. These organisms are usually introduced by exposure to environmental contamination. However,

[1]Deceased

another contamination source is the produced fluids arriving from the reservoir.

Until recently, investigators believed that only anerobic bacteria were found deep in the ground. And of these anaerobic bacteria, only two genera of sulfate-reducing bacteria (SRB), *Desulfovibrio* and *Desulfotomaculum*, were capable of causing corrosion problems in a terrestrial environment. Current research indicates that numerous aerobic, facultative anaerobic, and strict anaerobic bacteria do exist deep in the ground.[2] For example, geologists have determined that underground "rivers" circulate oxygen to extreme depths. At the same time, local areas are often cooled to the 30 to 50°C range. Under these conditions, even mesophyllic aerobic bacteria can grow at depths of several thousand feet. Furthermore, most U.S. production comes from reservoirs under secondary recovery. In these fields, surface water is injected into the producing formation to drive petroleum to a producing well.

Therefore, this injection water controls reservoir conditions for some distance away from the injection point. Bacteria growing in this injection water will contaminate the reservoir. These bacteria and their metabolic byproducts eventually contaminate the entire hydrocarbon production and water disposal system. Historically, this contamination cycle is completed in five to ten years after secondary recovery begins.

Common Oilfield MIC Symptoms

MIC within the petroleum industry appears to fall within two basic forms:

- Localized corrosion (usually pitting) under discrete biodeposits or "tubercles," and
- General corrosion at specific sites without accompanying biodeposits.

Tuberculation Corrosion

Localized corrosion with associated tubercles is called tuberculation corrosion. Tuberculation corrosion has been readily identified as MIC in other industries. However, within the oil industry, tuberculation corrosion occurring in anaerobic or reduced oxygenated systems has not been widely accepted as MIC. Yet the symptoms are identical to MIC found elsewhere:

- Iron-rich deposit materials that contain high levels of sulfides (and often phosphorus compounds),
- Iron-oxidizing and aerobic or facultative anaerobic slime-forming bacteria which are distributed throughout the deposit, and
- SRB at or near the metal surface.

These localized corrosion cells are found within piping and process equipment which contact water. Equipment operating between 30 and 50°C is particularly prone to biological contamination, and thus, to MIC. Furthermore, carbon steel is particularly prone to localized corrosion, especially corrosion centered around SRB. Carbon steel construction predominates within the industry.

General Corrosion at Specific Sites

General corrosion at specific sites is typified by corrosion and/or corrosion-fatigue failures of pump sucker rods. Oilfield corrosion specialists usually attribute these localized rod failures to corrosive reduction of rod cross-sectional area followed by corrosion-fatigue initiation. The corrosive medium is generally a mixture of water, carbon dioxide, hydrogen sulfide, and often minute amounts of oxygen.

Microorganisms may not be directly involved at these corrosion sites, but in many cases bacteria originally generated the CO_2 and H_2S. Currently, these corrosive gases are considered to be biologically produced when souring (production of hydrogen sulfide) occurs only after water injection was started.

Many oil and gas reservoirs contain naturally high levels of these corrosive gases. For these fields, it is unimportant whether these gases were originally biologically produced. These produced fluids are inherently corrosive and raise safety and toxicity concerns from the wellhead to the refinery (H_2S is more lethal to humans than hydrogen cyanide).

Dealing with Oilfield MIC

Tuberculation Corrosion

Tuberculation corrosion of aboveground piping and equipment is the most common oilfield MIC problem. To combat it, some companies inject biocides and/or corrosion inhibitors into the bottom of producing wells. Others inject such chemicals at key points in the aboveground process stream, particularly equipment handling the separated (produced) water.

While chemical treatments may reduce the problem, few companies claim to have completely stopped MIC by adding chemicals. The reasons for this are covered in the general discussion on biocides that follows. To adequately control MIC, a rigorous program combining physical cleaning with biocide treatment is necessary for troublesome systems. For example, frequent pigging, followed sequentially by high-level biocide and corrosion inhibitor treatments, is considered by many to be the best way to extend pipeline life. Pigging is not as successful once deep pits become established. Thus, these treatments must begin before serious problems develop. Unfortunately, most petroleum flow lines are neither amenable to pigging, nor are they treated with biocides.

Cathodic protection (CP) is widely used to protect submerged marine structures, buried pipelines, well casings, water-handling tanks, and other process equipment. Theoretically, CP offers a means of preventing problems where frequent physical cleaning is not practical. It has the added promise of arresting MIC already in progress, even within deep pits. However, CP has not been widely used to protect the interior surfaces of piping or well strings. Problems such as anode placement and maintenance will prevent this practice from gaining wide acceptance within the industry.

General Corrosion at Specific Sites

Filming amine corrosion inhibitors are commonly applied downhole to prevent general corrosion at specific sites. This application must be continuous for wells producing significant oil. Wells producing high volumes of water can be batch-treated at variable intervals (two to six months). The vast majority of the industry's anticorrosion expense is for downhole corrosion inhibitor applications.

Biocides and Oilfield MIC

The elimination of microbial contamination from process waters, particularly injection water, is tacitly accepted as a desirable goal within the industry. Unfortunately, controlling microbial contamination is difficult. In addition, controlling microbial contamination adds substantial costs to a nonrevenue-generating part of petroleum production: Water injection and disposal. As a result, substandard or no biocide treatment is the norm for the majority of U.S. production and injection systems.

Evaluating Biocides for Oilfield Use

An appropriate biocide, used properly, is required to combat the microorganisms involved in MIC. Traditionally, biocide selection methods within the industry are based upon the American Petroleum Institute's Recommended Practice No. 38 (API RP 38), the so-called "bottle test."[3] This method involves dosing water samples containing field bacteria with varying biocide concentrations. Viable cell numbers are then tracked with time to determine the optimum biocide dose rate to apply. Testing methodology, bacterial growth media, and incubation conditions are detailed in API RP 38.

Aside from inaccuracies inherent in using any one bacterial growth procedure for every oilfield situation, the API method only details testing for planktonic (free-floating) bacteria. Unfortunately, it is now known that the most troublesome microorganisms are those attached to surfaces (sessile), and sessile bacteria are fundamentally different from their planktonic counterparts. Planktonic methods are simply not appropriate for sessile bacteria. In addition, there may be no correlation between planktonic and sessile bacterial populations within a system. Hence, monitoring planktonic bacteria by traditional methods may not yield insight into sessile bacterial problems or MIC.

Evaluating Sessile Bacteria

Sessile bacterial analysis is inherently more difficult than planktonic analysis. For example, removable surfaces (coupons) must be in place for several weeks prior to evaluation. These coupons must be located at sites that truly represent the interior flow dynamics of the area being evaluated. Coupons must be carefully removed and the bacteria collected by nonstandardized methods. Finally, analyzing the bacteria present is usually done by standard planktonic methods which may not be appropriate for sessile bacteria.

Biocide Selection for Sessile Bacteria

Biocide treatments effective against planktonic organisms are not as effective against sessile organisms. Therefore, standard biocide screening procedures must be modified to properly analyze the chemicals' effects on sessile cells. Unfortunately, as with sessile analysis, sessile biocide screening tests are not standardized. Until standardized methods are issued, most biocide selection will continue to be done by outdated planktonic methods or by nonstandardized testing.

General Biocide Application Considerations

Other considerations besides kill efficiency must be evaluated when choosing biocides. One obvious consideration is biocide compatibility with other chemicals used within the system.

Another important consideration is the proper location of injection points. For example, chemicals that are compatible in use-dilutions may not be compatible if mixed in concentrated form. In addition, some biocides partition into the oil phase and can only be applied to separated fluids. Some biocides require special procedures to obtain a totally water-soluble chemical.

The effect of sulfide on biocides is of great concern. Sulfides react with biocides and severely reduce biocide effectiveness. Because these reaction products are not usually water-soluble, this creates problems with solids handling. Ensuring that the biocide does not interfere with filtration, ion exchange, or other processes is another concern.

Monitoring Biocide Applications and MIC

All biocide programs require proper monitoring for optimum effectiveness at the least expense. For MIC, this monitoring falls into two categories: monitoring a biocide program's effectiveness against sessile bacterial contamination and monitoring the program's ability to control MIC.

These tasks are not synonymous.

Bacterial populations must be routinely monitored, and trends must be recorded to successfully and economically manage a biocide program. Once done, biocidal effectiveness monitoring varies from a simple task for planktonic bacteria to a more difficult task for sessile bacteria. In either case, this monitoring can be done and a concrete result will be obtained.

However, monitoring biocidal effectiveness against MIC is not straightforward. It is not known how much microbial control is required to demonstrate that MIC has also been controlled. Therefore, some method for monitoring MIC directly must also be incorporated. This is usually done by monitoring corrosion rates in the system. Monitoring corrosion rates for oil field systems is complicated by the fact that MIC is seldom the only corrosion mechanism occurring.

Monitoring Corrosion Rates

During corrosion monitoring, field records on corrosion rates are normally used to establish a "base case" representing the untreated system. However, good historical records do not always exist. Therefore, it is necessary to establish a meaningful control, as with other biological problems, to prove that the applied treatment has an effect on the MIC. This is done in an identical parallel system, or by alternating chemical treatment with nontreatment to repeatedly observe treatment benefits. Neither procedure is widespread within the industry.

Summary

The reasons the petroleum industry suffers from MIC fall into three related categories:

- Industry operational practices,
- Industry failure to recognize the impact of MIC, and
- The lack of standardized methods to diagnose, control, and monitor MIC.

These reasons are expanded below.

Operational Practices

The petroleum industry does not routinely act to prevent microbial contamination or to reduce biofouling within their systems. This is primarily because money spent on a nonrevenue-generating aspect of the business is difficult to justify. This is particularly true because a cost is never placed on failing to prevent microbial problems.

Industry Ignorance About MIC

MIC has not been widely accepted as a major source of corrosion problems within the petroleum industry. The impact of MIC is unproven, as is the inherent benefit of preventing bacteria contamination. Sufficient data are not available to show how much "general corrosion" is actually MIC within the industry. Even less data are available to show that attempting to control microbial contamination will reduce overall corrosion costs.

Lack of Standardized Methods

To assess MIC's true impact on the industry, it is necessary to gather sufficient data. These data cannot be gathered efficiently nor can different situations be compared until standardized data collection methods are accepted and used. The American Petroleum Institute has recognized this need and is rewriting RP 38 to address sessile bacterial monitoring methods. However, methods for directly monitoring MIC will not be developed. These methods are critical for the industry to effectively control MIC losses.

References

1. K.C. Lunden, T.M. Stastny, "Sulfate Reducing Bacteria in Oil and Gas Production," CORROSION/85, paper no. 296 (Houston, TX: NACE, 1985).
2. R.M. Atlas, ed. *Petroleum Microbiology* (New York, NY: Macmillan Publishing Co., 1984).
3. "Recommended Practice for Biological Analysis of Subsurface Injection Waters," API RP 38, 4th ed., (Washington, DC: American Petroleum Institute, 1982).

MIC in Marine Industries

By R.G.J. Edyvean and S.C. Dexter

Introduction

Most instances in which biological organisms have influenced corrosion in the marine environment can be illustrated by the experience of the offshore oil and gas production industry. Offshore oil and gas production "rigs" are large and complex structures, consisting of a subsea "jacket" and a topside production platform. The jacket, which may stand in several hundred feet of water, is usually made of steel tubing, ranging in diameter from 30 to 550 cm, with wall thicknesses of up to 15 cm. Above sea level, there is 30 to 50 m more of jacket structure supporting the platform. This is the site of the oil, gas, and water handling plant, drilling operations and living accommodations. The hollow legs of the jacket may be concrete- filled at their bases or flooded with seawater. Some concrete structures may use the legs and other seabed cells to store oil.

When the production of oil and gas from shallow offshore platforms began in the 1950s in the Gulf of Mexico and in the mid-1960s in the North Sea, the possibility of biological problems was overlooked for many reasons. First, such problems usually take time to become apparent. Second, the effects of marine fouling, bacterial corrosion, and bacterial hydrogen sulfide production were underestimated. Third, the industry was busy with initial production operations, and set patterns of production, storage and transport had not yet been established. Finally, the extent of the reserves, and hence the longevity, of the production facilities was unknown. Also, the producers had a certain competitive reluctance to be the first to acknowledge any problems.

By the late 1980s, however, offshore oil and gas production had been established for many years and in many locations including the North Atlantic, North Sea, Black Sea, Persian Gulf, Indian Ocean, China Sea, Tasman Sea, South Atlantic, and Caribbean, as well as the original Gulf of Mexico sites. Recently, a considerable amount of information on biological corrosion problems offshore has been released.[1-4]

The offshore industry has all the problems of onshore plants where hydrocarbons and water are present, and it has the added problem of being situated in seawater, which is a highly corrosive and biologically active medium. It is this presence of water, essential to all biological processes, which is the most important factor in all biological corrosion problems.[5] The influence of biological organisms falls into three main interactive classes: Fouling (coverage and blockage by both macroorganisms and microorganisms); souring, due to hydrogen sulfide production by microorganisms in oil or water; and accelerated corrosion, due to an influence of microorganisms on corrosion electrochemistry.

There is also an important distinction between the different environments encountered offshore and the type of biological activity in each. One environment is external surfaces, exposed to seawater, and marine fouling ranging from bacteria to macroalgae several meters long. The second is the interior surfaces of plant and pipework, which can be conveniently subdivided into the seawater handling and oil production systems. These internal surfaces are fouled mainly by bacteria. Figure 7.1 summarizes the areas of an offshore platform where biological corrosion problems are known to occur. The following review deals with each area separately. Similar problems are also experienced (although less well-documented) on other off-

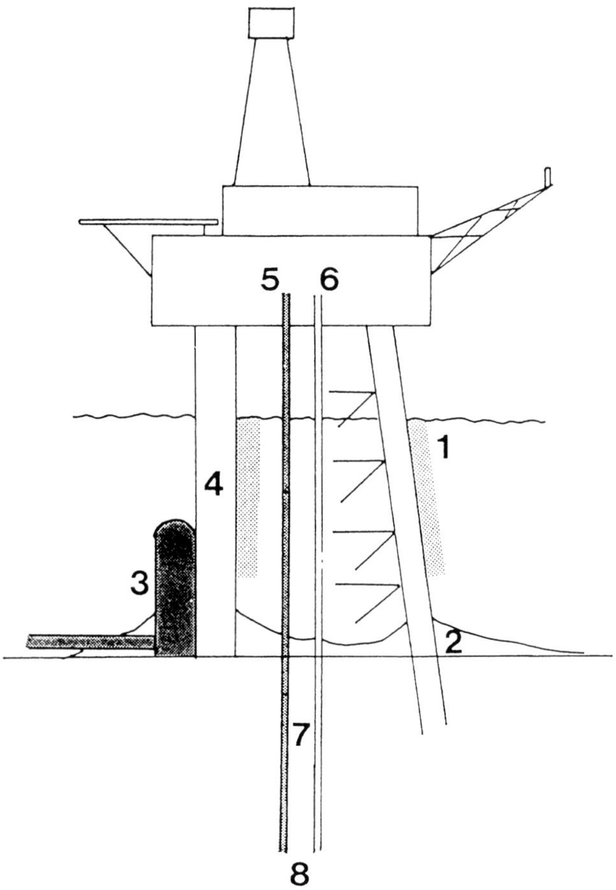

FIGURE 7.1
Composite diagram of an offshore structure showing the main sites of biodeterioration problems: 1—marine fouling; 2—drill cuttings around legs; 3—oil storage and transport; 4—water-filled legs; 5—production system; 6—seawater injection system; 7—downhole pipework; 8—reservoir problems. (After Sanders and Hamilton,[20] and a poster produced by SGS Micran Ltd.)

shore and coastal structures such as shipping, port facilities, and power plants.

External Effects

The external surfaces of offshore platforms are exposed to a range of environments from the highly aerated "splash zone" influenced by tides, currents, wind, and weather to the highly anaerobic sediments at the base of the structure, influenced by drilling mud and other material dumped from the rig, as well as the chemical and organic nature of the sediments themselves. A considerable number of sessile and sedentary marine organisms, both microscopic and macroscopic, can readily colonize any surface placed in the sea. These organisms are known as marine fouling, and together with accumulated debris they can build up to a significant thickness (tens of centimeters) on the metal surface.

Sequence, Occurrence of Fouling

The sequence of fouling of a surface placed in seawater is well-documented.[6-9] Initial chemical interactions between the material and the seawater result in the adsorption of dissolved organic material onto the surface, imparting an ionic charge.[10] While these films may or may not vary on different surfaces, they generally facilitate the settlement of bacteria and microalgae, such as diatoms.[11,12]

Bacteria and diatoms can be present in large numbers soon after immersion of a surface, and they produce a "slime" that can reach 2 mm in thickness and contain approximately 30 million bacteria gm^{-1} [13] and 270,000 diatoms cm^{-2}.[14] Following the establishment of this microcommunity, macrofouling gradually develops. In engineering terms this community comprises "soft fouling," such as: algae, soft corals, sponges, anemones, tunicates, and hydroids; and "hard fouling," such as: crustose algae, barnacles, mussels, and tubeworms.

It should be remembered that the fouling community is dynamic, and individual species of both macrofouling and microfouling are competing for space and dominance. Thus, at any one time, different parts of the structure may have a slime film, soft fouling, or hard fouling on both large (several meters) and small (centimeters) scales.

Marine fouling on offshore structures in the North Sea between the United Kingdom and Norway has been summarized in several recent publications.[15,8] It has been found that the composition, amount, and rate of development of this fouling depends to some extent on the location (geographic, depth, and aspect) of the platform and season of its immersion.[16] An excellent review of marine fouling can be found in the two books: *Marine Fouling and its Prevention* [6] and the sequel, *Marine Biodeterioration*.[7]

While bacteria are ubiquitous in seawater, larvae and the spores of larger organisms will arrive at structures by water currents from coastal and nearby platform populations, and from fouling on passing ships. Different types of communities, characteristic of different depths, can be recognized on many platforms with mussels and algae nearest to the water surface, soft corals, anemones and hydroids lower down, and tubeworms, barnacles, tunicates, hydroids and soft corals at greater depths. When planning for offshore structures, it is important to determine the nature and growth rate of fouling in the particular region of interest. As a rule of thumb, high levels of fouling should be expected in areas of high fisheries' production and in areas with cold, nutrient-rich waters, well supplied with oxygen and sources of organisms in their settling stages.

The debris from drilling operations is usually dumped around the base of the platform. These cuttings are mixed with thixotropic drilling "muds," which are used to lubricate and lift debris from the area of the drill head. These muds can be a complex mix of organic and inorganic chemicals, weighting and gelling agents.[17,18] While most drilling muds are recovered from the cuttings prior to dumping, a substantial proportion remains. The buildup of these muds can result in considerable microbial activity around the legs of structures at the critical water/seabed interface.[19,20]

Effects of Fouling

The general ways in which marine fouling can damage offshore structures are: obscuring the substratum, enhancing corrosion, increasing hydrodynamic loading, and enhancing corrosion-fatigue. These will be discussed in the following sections.

Obscuring the Substratum

Covering the surface of the structure with a heavy fouling layer reduces the ease and accuracy of either direct or remote inspection of the metal surface, thus increasing the probability of surface corrosion or cracks going unnoticed. This results in the need for fouling to be cleared before visual inspection and nondestructive testing can be carried out. The amount of time spent clearing fouling increases the costs and risks of the inspection process considerably.[21,22]

The partial barrier that marine fouling places between the seawater and the metal surface of the submerged structure allows a considerable range of macrobiological and microbiological environments to form. Marine macrofouling grows in intimate association with bacteria, and the interactions between the two are of major importance in seawater environments. Thus, macrofouling can affect corrosion processes, both directly and indirectly, by modification of the microenvironment (including bacterial activity) at the metal surface.

The close interactions found between different species of bacteria extend at least to bacterial-microalgae interactions.

Also, the continual cycle of growth and decay of the macrofouling is a major source of nutrients for bacteria.

One result of the macrofouling layer is the production of oxygen-deficient environments at the metal surface suitable for growth of the anaerobic sulfate-reducing bacteria (SRB). The SRB subsequently produce hydrogen sulfide (H_2S), which creates a "sour" environment, even though the bulk of the seawater is well-aerated. Fouling has also been found to cause considerable increases in loading due to wave action, thus enhancing the destructive forces of corrosion fatigue. The effects of SRB and increased loading will be discussed in subsequent sections.

Enhancing Corrosion

Seawater is a well-known corrosive environment, and while there is some evidence that a uniform layer of fouling can reduce the rate of corrosion by excluding oxygen, biological activity is rarely uniform and tends to enhance the aggressive nature of seawater.[23] Fouling, especially by bacteria, which are in the most direct contact with

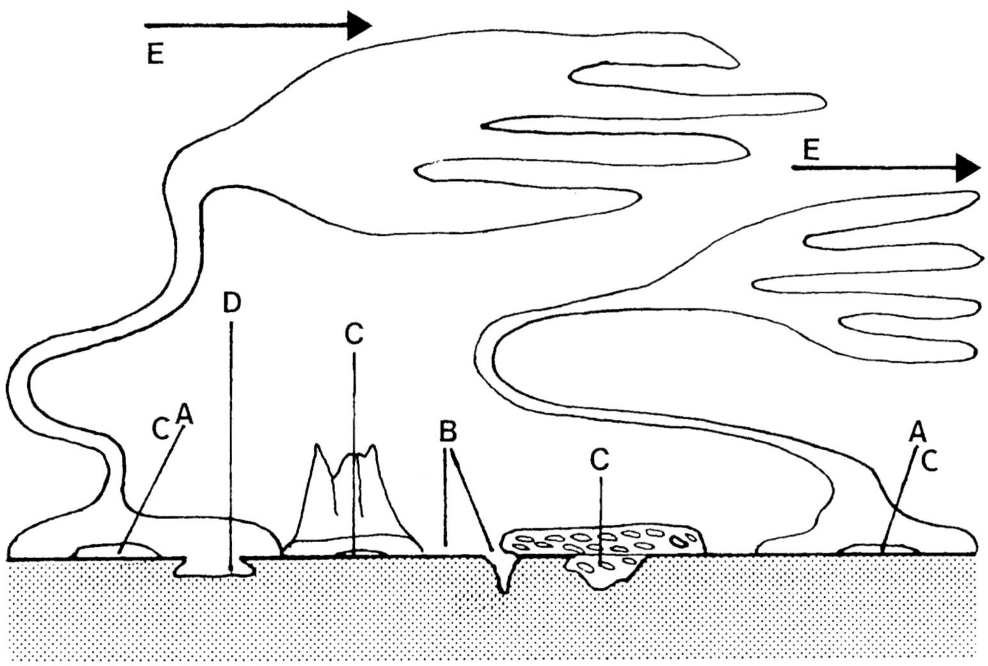

FIGURE 7.2
Influences of fouling on offshore structures. A—areas of trapped water. B—Oxygen concentration cells. C—Sites for anaerobic bacteria. D—Removal of bright metal. E—Drag effects-fatigue.

the metal surface, will influence the corrosion of metals in several ways:

- By their physical presence creating differential aeration, pH or other concentration cells on both macroscopic and microscopic scales,
- By the production of corrosion-promoting metabolites such as acids (both organic and inorganic) and sulfur compounds (especially H_2S),
- By stimulation of cathodic (reduction) reactions, and
- By stimulation of anodic (oxidation) reactions.

Several reviews have been published on the enhancement of corrosion by:

- Marine fouling,[24,25,9]
- Algae,[26,27]
- Sulfate-reducing and sulfur-oxidizing bacteria,[28,29] and
- Aerobic biofilms.[30,31]

The general mechanisms by which biological organisms enhance corrosion occurring on the external surfaces of offshore structures are the same as those for any other biocorrosion situation. Figure 7.2 shows some of the influences of fouling on a metal in seawater. In areas of trapped water (point A in Figure 7.2) and crevices (points B and C in Figure 7.2) under either macrofouling or bacterial biofilms, considerable changes in pH, dissolved oxygen, and ionic balance can occur. These changes can enhance corrosion by the formation of differential pH, aeration, and chemical concentration cells. The daily pH change under healthy algae, due to their photosynthetic action, typically exceed 2 pH units from a baseline of 8 to over 10, and when the buffering capacity is broken, the pH can fall to approximately 6 during the dark.[32] Levels of pH between 4.5 and 7.5 have been measured beneath animal fouling[33] and, if bacterial acid decomposition takes place, pH levels as low as 1.8 can be created.[34] Thus, the total difference in pH between two adjacent sites can be up to 9 pH units, and this can form highly corrosive concentration cells.

Oxygen concentration cells (point B in Figure 7.2) can lead to pitting and crevice corrosion. These forms of corrosion are very important in seawater environments where extreme differences in oxygen concentration can be created between portions of bare metal surface in contact with oxygenated water and other areas of metal covered by fouling or even thin films of bacteria.

Corrosion by the metabolic products of bacteria and the direct or indirect stimulation of anodic and cathodic reactions are similar to those in other biocorrosion situations. A wide range of potentially corrosive metabolites are produced by both macroorganisms and microorganisms. Some of these are released directly into the seawater away from the metal surface, and others accumulate at the surface within the biofilm. So, while macroalgae such as *Desmarestia sp.* can produce sulfuric acid at a pH of 1.8,[35] this may or may not cause a problem.

Enhancement of the cathodic reaction on passive alloys by the action of general aerobic bacterial films has been well-documented.[36-38] The effect occurs within approximately two weeks of immersion in most natural waters of varying salinities.[39-41] In some waters where thick biofilms form rapidly the enhancement is minimal or absent while in other waters where biofilms form slowly the effect can take as long as one year to become significant.[42,43] When this effect occurs, it leads to accelerated initiation and propagation of localized corrosion, usually at crevices, on susceptible stainless steels,[44] and it increases the severity of galvanic corrosion whenever electrochemical reactions on the cathodic member of the couple are rate controlling.[45,46]

Enhancement of the corrosion reaction through the production of H_2S by the SRB has also been widely reported[47] while anodic enhancement has been demonstrated for nitrate-reducing bacteria[48] and blue-green algae.[49] However, H_2S production is still an unknown factor on external surfaces offshore. High SRB activity has been detected,[24] but our knowledge of H_2S levels near the metal surface and their effects on hydrogen entry into the metal, in turn affecting hydrogen embrittlement and corrosion-fatigue, are still reliant on laboratory simulation. These aspects are discussed further in the next section.

Hydrodynamic Loading and Corrosion Fatigue

Fixed offshore structures are susceptible to two main deleterious forces, the corrosive environment as modified by biological action and cyclic stress loading due to wave action. The first 30 m of the structure below the water line carry some 70 percent of the total loading due to wave action,[50] and the presence of fouling, especially large algae (e.g., kelps), that dominate this zone in North Sea environments, will considerably enhance this loading. Marine fouling can enhance the fatigue stresses (cyclic loading from waves and tides) on an offshore structure by increasing both the diameter and roughness of tubular members of the structure. Both of these factors increase the drag coefficient and, hence, the drag forces on the structure.[51,52]

It has been calculated that a layer of fouling 150 mm thick can increase the hydrodynamic loading by as much as 42.5% and the fatigue damage by 62%.[50] Heaf also calculated that this would decrease the fatigue life (the predicted number of stress cycles to failure) by 54%.

Recent studies summarized by Vaux[53] (Figure 7.3) indicate that the force coefficient can be markedly increased by the presence of kelp fouling. The tests carried out at the Christchurch Bay Tower (CBT) showed up to a 100% increase in the drag coefficient for a vertical cylinder.[53] Other studies (Figure 7.3) also found higher forces due to fouling. Studies by Theophanatos and Wolfram[54] indicate that hard fouling, such as mussels and barnacles, will increase the drag factor by 1.5 times over a macroscopically smooth cylinder while large frondose algae, several meters long, can increase the loading by as much as three times.

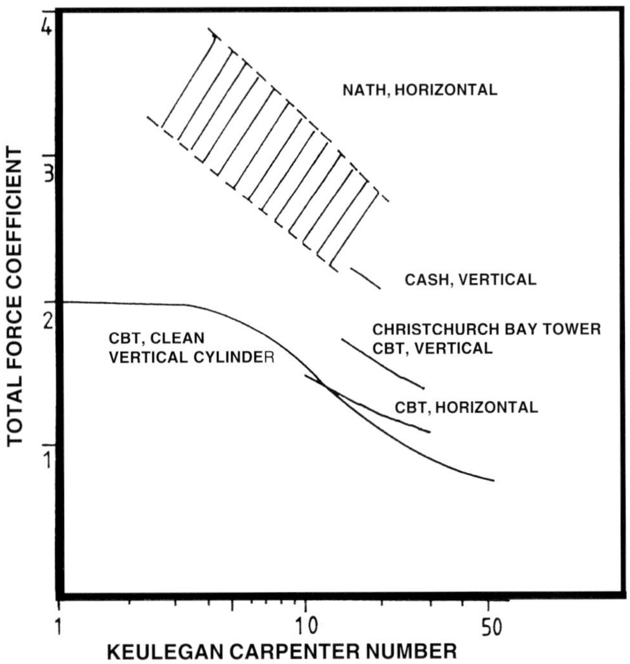

FIGURE 7.3

Total force coefficients with macroalgae (kelp-real or artificial) on cylinders from different experiments compared with a macroscopically smooth cylinder (CBT clean vertical cylinder) CBT—Christchurch Bay Tower. CASH—National Physical Laboratory results. NATH—Oregon State University results.[53] (Crown copyright).

A corrosive environment and cyclic stress on a structure combine to produce the phenomenon known as corrosion fatigue. This causes failure by the formation and growth of cracks. Corrosion fatigue develops in three stages: crack initiation, crack growth, and final catastrophic failure. Processes, such as formation of concentration cells and the production of aggressive metabolites, may lead to crevice and pitting corrosion, which, in turn, can initiate fatigue cracks.[28] However, these processes are not necessary for crack initiation because complex welded structures, such as offshore production platforms, always contain flaws that would reduce or eliminate crack initiation time. Thus, the rate of crack growth is the most important parameter in designing for and predicting the life of the structure.

The rate of crack growth under corrosion fatigue conditions is controlled by dissolution of metal at the crack tip and hydrogen embrittlement.[55] Hydrogen embrittlement involves the entry of atomic hydrogen into the metal, causing a loss of toughness, thus lowering the loads required to cause fracture. Any external supply of hydrogen to the metal is likely to increase hydrogen embrittlement and crack growth rates.

H_2S is particularly aggressive because the sulfide ions stimulate anodic dissolution and inhibit the combination of hydrogen atoms to form gaseous molecules, thus making the hydrogen atoms more readily available to enter the metal and cause embrittlement. When it was realized that biologically generated hydrogen sulfide could affect the main structure, as well as the pipelines and production plant, measurement of corrosion fatigue crack growth rates in such environments became necessary.[56]

Corrosion fatigue data are derived from tests in which the crack growth rate is measured for precracked specimens of the appropriate structural material subjected to known levels of cyclic stress in a corrosive environment. By using carefully designed specimens, crack growth rates measured on the specimen can be directly related to the structure by use of the crack tip stress intensity factor, K.[57] For any metal, there is a value of K, termed K_{1C}, at which catastrophic failure occurs. In a corrosive environment, a load which produces a value of K less than K_{1C} may still cause crack growth by a combination of mechanical and chemical factors, the rate being a function of the environment. There is also a value of K below which no crack growth will be observed.

Under cyclic loading, the stress intensity at the crack tip will vary as the crack loads and unloads, and the range of stress intensity in each cycle is denoted ΔK. In corrosion-fatigue studies, it is this parameter (ΔK) against which the instantaneous crack growth rate, da/dn, is compared (see Figure 7.4), where a is the distance (meters) that the crack grows per loading cycle, n. For a fixed load range and specimen geometry, ΔK changes in proportion to crack length. This means that one specimen will give crack growth data for the whole range of ΔK up to catastrophic failure.

In early work to determine the effect of H_2S on

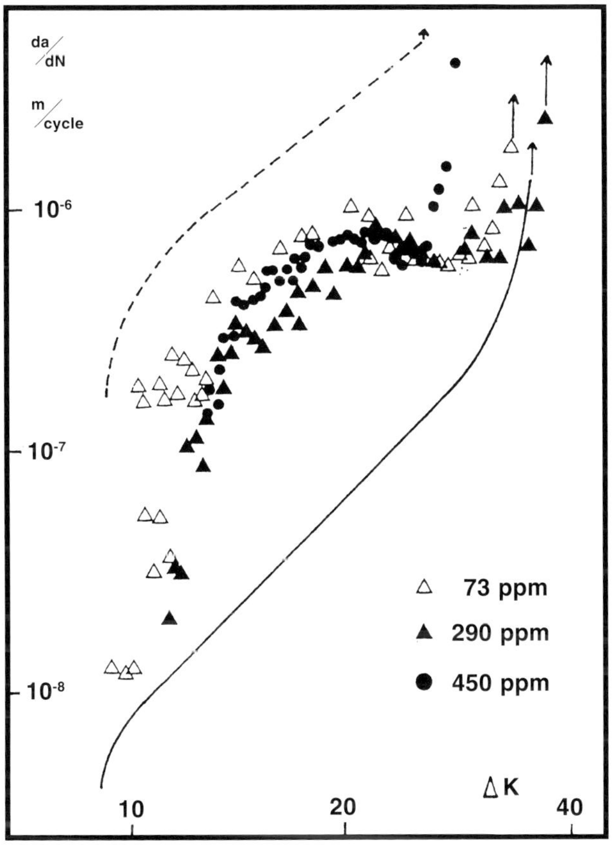

FIGURE 7.4
Crack growth rates of a RQT 501 steel in biologically active H_2S seawater environments; solid line—crack growth rate in seawater; dotted line—crack growth rate in 520 ppm H_2S in abiotic natural seawater.[57] (Reprinted with permission of Harwood Academic Publishers GmgH.)

corrosion fatigue in seawater, saturated aqueous solutions were used, but these were found to be more severe than either the sour oil or marine fouling conditions.[58]

More recent work has concentrated on testing at the levels of H_2S that could reasonably be expected in offshore environments. Reports of H_2S levels due to bacterial action vary from 15 to 100 ppm in the interior of platform legs[59] to more than 2,000 ppm in pure cultures.[60] Levels associated with bacterial films are likely to range from 10 to 50 ppm under purely bacterial biofilms to 300 to 400 ppm where macrofouling decomposition produces a highly active bacterial culture.

Thomas et. al.[57] used seawater in which various levels of H_2S were generated by allowing the natural bacterial decomposition of macroalgae. Their results (Figure 7.4) show that crack growth rates are 10 to 20 times higher than those in seawater, even at low levels of H_2S. They also showed that there is a "plateau level" of biologically produced H_2S, above which there is little increased effect on crack growth rates, except at high values of K, and this maximum level is less than that produced by similar levels of abiotically generated H_2S. These results indicated that a biologically active environment is markedly different in its effects from a simple solution of H_2S in seawater, and that further study is required in this area.

For H_2S production to have an effect on corrosion fatigue of offshore platforms, three factors have to coincide: a suitable level of H_2S, a suitable point for stress concentration on the structure, and a preexisting crack. Perhaps the most significant factor is the coincidence of a crack with the enhanced corrosive environment. Crack initiation can be a long process in corrosion fatigue. Thus, as a structure gets older there is a greater probability of cracks occurring at high stress points, and there is also a greater probability of H_2S generation, an important point when deciding on an inspection policy for older structures.

Seabed Problems

At the seabed, oxygen availability tends to be lower, fouling growth is less, and cyclic loading not as marked. However, the water/mud (or sand) junction provides possibilities for bacterially assisted concentration cells, and corrosion problems may be enhanced by debris from the platform, especially from drilling operations. The drilling cuttings are mixed with thixotropic drilling "muds" used to lubricate and lift debris from the area of the drill head. The buildup of these muds can result in considerable microbial activity around the legs of the structure at the critical water/seabed interface with subsequent corrosion problems.

Protection Against Corrosion

The submerged areas of offshore structures are protected against corrosion by cathodic protection combined, in some cases, with protective coatings. The effect of cathodic protection on corrosion fatigue is ambiguous, having been shown to be either detrimental or protective, depending on conditions.[58] Thus, it should be noted that cathodic protection alone may not be the complete remedy for corrosion fatigue problems. For example, it has been observed[61] that SRB can be as active on cathodically protected as on freely corroding steel.[20] Thus, care in the monitoring and maintenance of both the structure and cathodic protection systems is essential.

Recently, there has been an increased interest in antifouling coatings. These include both the conventional paints and the newer composite systems such as copper mesh mounted on a neoprene backing. These "retrofit" systems are expensive and tend to be used only where essential. Antifouling coatings are difficult to maintain on offshore structures because of the lack of periodic dry docking, but paints and other antifouling systems are being developed for use in sensitive areas. Other antifouling measures that have been considered include the use of current-driven discs rotating around the tubular members of the structure and even encouraging predation of mussels by starfish.[62]

The presence of marine fouling can lead to physical damage to structures and auxiliary equipment, particularly when the fouling is removed by storm action or cleaning processes. Physical damage includes:

- Removal of paint coatings along with the fouling,
- Damage to cathodic protection cables,
- Covering and functional impairment of sacrificial anodes used to protect the structure,[62,63]
- Removal of coatings and corrosion products on bright metal when holdfasts of large kelp are pulled away from the surface.

Biological activity can enhance corrosion even when cathodic protection is applied. It is known that the presence of SRB necessitates a shift in the cathodic protection potential to compensate for their activity, but this is not always easy to achieve in affected areas of the structure while, at the same time, avoiding overprotection. For example, pitting corrosion and the presence of iron sulfides, evidence for the action of SRB, have been observed on some platforms in the North Sea despite the use of cathodic protection.[61] This was attributed to cathodic protection potentials not being negative enough to ensure adequate protection. Areas such as flooded legs and oil storage tanks can be periodically sparged with air or seawater. While this is effective against hydrogen sulfide production, it may increase corrosion by the production of oxidizing conditions and the encouragement of acid-producing bacteria.

Hardy and Bown[64] noted that SRB corrosion was most severe when anaerobic and aerobic conditions were alternated. These observations point to the need for awareness of the potential problems, good design, adequate monitoring and proper maintenance.

Internal Effects

The two areas of platform operation in which most biodeterioration problems occur are the seawater handling system and the oil production system. Seawater has many uses offshore, but by far the greatest use is for secondary recovery, whereby water is injected into the reservoir to force oil to the production wells and maintain reservoir pressure. The oil production system often has a high water content due to seawater injection or from water naturally present in the reservoir. Apart from the seawater filtration system, nearly all problems in these two areas arise from bacterial growth fouling the surfaces of pipes and plant, causing blockage, decreasing heat exchange efficiency, and enhancing corrosion and "souring" the oil or water with hydrogen sulfide. Thus, before examining the two areas themselves, the types of bacteria involved will be described.

Bacteria Involved

There are several groups of microorganisms that will thrive in both areas. The main groups are:

- SRB,
- Iron bacteria,
- Slime-forming bacteria,
- Sulfur-oxidizing bacteria,
- Nitrogen-utilizing bacteria,
- Fungi, and
- Plankton (in seawater injection systems only).

SRB are of particular concern. Although described as obligate anaerobes, many SRB species have the ability to survive as "resting stages" or spore forms in oxygenated seawater. They are easily introduced into water injection and oil production systems, and they become active in any area of low oxygen concentration, such as under biofilms or deposits. The sulfide produced may react with solubilized iron to form FeS. This, in combination with bacterial cell masses, is an excellent plugging agent and may block oil-bearing strata and clog filters, valves, and other control equipment.[65-67] Furthermore, since FeS is cathodic to steel, it may cause increased pitting corrosion. Additionally, hydrogen sulfide (H_2S), a corrosive agent in its own right, is produced. H_2S "sours" oil and gas, and this not only reduces the value of the product,[66,68] it also extends the corrosive effects along pipelines. SRB can also enhance corrosion directly by enzymatic stimulation at the metal surfaces.[47]

Iron bacteria, such as *Gallionella* and *Crenothrix,* have been reported to cause a number of problems in oil field systems.[69] They are characterized by "cotton wool-like" colonies with the bacteria surrounded by sheaths of slime. Often found in association with SRB, the slimy precipitates of hydrated ferric oxide that are formed will reduce filter efficiency and, if they reach the reservoir, may decrease the permeability of the reservoir rock. Although the major problem associated with iron bacteria is plugging, their growth may also result in decreased pitting corrosion by differential aeration and acid ferric chloride production.

Slime-forming bacteria are not a single group of organisms. Many bacteria can produce a slimy covering given the right environmental conditions (e.g., the iron bacteria above).[70,71] They cause problems by plugging filters, injection lines, and reservoir rock formations. Formation plugging can be a serious problem because remedial measures, such as acid or surfactant treatment, are difficult to apply and not always effective. Dense slime masses adhering to the sides of lines and equipment will exclude oxygen from the metal surface, resulting in differential aeration cells and conditions conducive to the growth of SRB. Slime also reduces efficiency of heat exchangers and may degrade hydrocarbons.[72]

Sulfur-oxidizing bacteria oxidize sulfur compounds, such as sulfide, to sulfate or sulfur. They are often found in association with SRB where they can form corrosive "sulfureta," that is, alternating oxidized and reduced sulfur environments.[73] In enclosed systems, the oxidation of hydrogen sulfide and other compounds to sulfate can form highly corrosive acid conditions.[28,74] The optimal temperature for species of *Thiobacillus* and *Sulfolobus* can range up to 90°C, oxygen concentration from zero to saturated, and pH from neutral to very acid with a pH optimum for acidophiles of 2.5.[74]

Nitrogen-utilizing bacteria are involved in the cycling of nitrogen compounds. Organisms such as *Nitrosomonas* and *Nitrobacter* will oxidize ammonium to nitrite and nitrates respectively while nitrate is used by anaerobic nitrate-reducing bacteria such as some species of *Pseudomonas.* The production of acids, stimulation of anodic dissolution,[48] and formation of consortia (or communities) of different bacteria will enhance the risk of corrosion. The above bacteria have been shown to be significant in enhancing corrosion in oil pipelines, especially those carrying oil from fields using water floods.[75]

Fungi are usually aerobic, and they can generate organic acids and other metabolites to promote corrosion. They can use hydrocarbons as nutrients and can be a problem in fuels.[76] Plankton, microscopic plants and animals floating in seawater, are of significance in seawater injection systems where they can cause extensive blockage of filters and provide nutrients for bacterial growth.[77]

MIC has been described in many areas within the North Sea Oil industry, and almost all aspects of oil treatment can be damaged by the effects of microbial activity.[20] Corrosion damage may result from direct or indirect action by microorganisms. Direct action from the organisms growing on the metal surface, as discrete colonies or as slime films, will produce effects such as differential aeration or the production of iron sulfides, resulting in pitting attack which can become self-propagating.[78] Indirect action can result from the growth of organisms remote from the corroding metal surfaces. Examples are the production of acids and sulfides ("sour" conditions), causing corrosion problems downstream of the areas of growth.

However, in almost all cases, the biological influence results in localized rather than uniform attack.[79] Specific effects on the oil production and water handling systems are described below.

Production Systems

The operation of oil production systems offshore can create ideal conditions for bacterial growth, especially SRB, with severe corrosion of production equipment.[20] The production system consists of a series of separators, gas scrubbers, condensers, coolers, and pumps that separate the oil, gas and water, dewater the oils; and clean and compress the gas.

Biological corrosion problems are most likely to occur in the production system when seawater has been introduced by the injection wells. As the reservoir is gradually depleted of oil, the amount of water in the oil will increase, as will its biological activity.[19] It should also be remembered that this situation will often be at a stage when the production equipment has been in use for some years. The resulting emulsion of oil, water, and gas, and the plant through which it passes can be subject to severe biodeterioration problems, especially "souring" and corrosion.

Once the oil has been cleansed of the water and gas, it has to be stored, awaiting transfer to a tanker or transported to shore through pipelines. Both these components have biodeterioration problems similar to the rest of the system. Oil storage can cause considerable problems, especially because the tanks are often used to ballast the structures and, thus, have to contain seawater when not holding oil.[73] This water rapidly becomes stagnant, and significant concentrations of hydrogen sulfide can be generated and released into the gas space above the water.[59] This can cause a serious health hazard and may also produce hydrogen embrittlement of metal exposed to the gas space. Water displacement oil storage systems are used in conjunction with skimmer tanks and oil recovery tanks to separate any oil droplets from the displaced water.

Again, the whole system is an excellent breeding ground for bacteria, and both SRB and aerobic hydrocarbon-oxidizing bacteria have been shown to coexist.[59,80] The hydrocarbon-oxidizing bacteria degrade oil and use up available oxygen, thus providing both the nutrients and conditions for SRB activity.[73] Once H_2S has been formed, or other reduced sulfur compounds are present, then oxidation can produce sulfuric acid.[74]

In addition, Westlake et. al. have found relatively large populations of aerobic, facultative, and anaerobic bacteria in fluids and tubercles in oil-carrying pipelines.[75] This is especially common in pipelines serving fields using water injection procedures. These authors attribute corrosion enhancement by anodic depolarization (stimulation of the metal-metal ion reaction) to these bacteria.

Bacteria are highly tolerant of changes in environmental conditions, and some are adapted to growth at extremes of temperature, pH, and pressure. Thus, high temperatures in the production system will not necessarily prevent bacterial activity especially as there are likely to be "cool spots" in any hot system and "hot spots" in any cold one.[5] It should be remembered that the corrosion rate is also markedly influenced by temperature, doubling for every 10°C.[81]

Seawater Systems

Seawater injection is now used in nearly all offshore production systems as better oil recovery is achieved by maintaining a constant pressure in the reservoir. Typical injection wells are 11,000 to 12,000 feet deep and supply 50,000 barrels of water per day at a pressure between 3,500 and 4,500 psi.[82] Complete systems handle up to 500,000 barrels (80,000 m^3) of water per day, using several injection wells.

However, seawater is corrosive and carries bacteria, plankton (microscopic plants and animals), and other contaminants that can cause blockage of the well and reservoir and enhance corrosion.[83] To remove these contaminants, the water is filtered (usually with the goal of removing 98 percent of all particles above 2 microns,[77] but considerable problems have been found with the filtration process.[84] It has been reported that more than 90 percent of suspended matter in seawater is organic, with plankton making up a large proportion of the particles.[85] The proportion can change rapidly during seasonal growth periods ("blooms") of plankton.[86]

The extensive pipework in the seawater system has to be kept free from corrosion and fouling. Even though biocides are added and the water is chlorinated and deaerated, corrosion problems can still arise, particularly if there are incompatibilities between the biocide and corrosion inhibitor. Thus, the seawater system can experience biofouling and corrosion problems in the water intake, filtration equipment, subsequent pipework, and plant. In the water uptake, growth of large fouling organisms, such as mussels, can restrict or even plug intake lines, and will increase the solids loading of the water.[87]

In the filtration system, the growth of bacteria, algae or other organisms on or within the filter can cause considerable impairment of filtration efficiency. Dead algae, fungi, and oil are all conducive to bacteria growth in the filter. Without proper cleaning and sterilization, filters, especially deep bed filters, become black from the activity of SRB and other bacteria.[88] Such growths in the filter media can form solid masses that eventually destroy the filtering capacity.[89,90] Eventual blocking of filters by organics and particulates from the seawater is almost inevitable, whether the system consists of disposable or backwashable elements. However, care in design and operational factors can considerably lengthen filter life or backwashing cycles.

Injection systems, as presently designed, can provide ideal conditions for the growth of SRB and other bacteria, thus leading to corrosion problems.[5] Suitable sites for growth include tanks, bends, and drains, etc. Such sites often have large amounts of sulfate, low oxygen, elevated temperature due to heat exchangers and high nutrient levels from filtration and fuel gas strippers. Case has found that most corrosion failures in oil field water systems are due to hydrogen sulfide (H_2S) and oxygen in about equal proportion, followed by "other" bacterial problems.[91] Oxygen is very important in promoting corrosion, and it is usually removed from the seawater using gas stripping or vacuum deaeration and chemical scavengers to leave a residual oxygen content of 0.1 ppm or less. Prior to deoxygenation, corrosion resistant or nonmetallic pipework is required.

However, deoxygenation is not a cure-all for corrosion problems, and deoxygenation of a system still rich in organic material can provide ideal conditions for anaerobic bacteria. Thus, suitable corrosion inhibitors and biocides are needed. As most H_2S is produced by bacteria, their activity is a very important factor. Corrosion rates of up to 1 mm per year have been measured together with significant pitting. Studies by Herbert[92] and Herbert and Stott[93] have also indicated that the pressures encountered in water-injection systems (approximately 4,000 psi) are unlikely to offer a barrier to the growth of SRB.

The attached sessile bacterial population is more numerous and important than the planktonic one, and misleading results can be obtained if this factor is overlooked.[20] For example, SRB can be found in large numbers in biofilms on pipewalls when they are in very low numbers in the water.[5]

Stagnant zones beneath scale or other deposits, including biological slimes, and throughout the system downstream of deaerators will provide anaerobic sites for SRB activity.[87] Unless controlled, these bacteria can reach the reservoir and turn sweet produced fluids sour on breakthrough when the injected water reaches the production wells. Carlberg noted that virtually all seawater floods eventually become sour.[87]

Reports of H_2S corrosion in water systems are quite variable and not proportional to measurable H_2S concentrations. Case observed that corrosion is usually mild in clean systems regardless of the amount of H_2S present.[88] However, in fouled systems, corrosion could be severe despite very little detectable H_2S in the water. He ascribed this to the formation of a thin and protective film of corrosion product in clean systems while in fouled systems bacterial activity under deposits produced deep pits filled with sulfide corrosion product.

Control Measures

In internal systems, the main control measures involve biocides. The past ten years have seen considerable advances in the understanding, testing, monitoring, and control of biological corrosion problems offshore. While internal water injection systems have been shown to be very susceptible to biologically induced problems,[20] it also has been shown that suitable monitoring and treatment measures can result in effective control.[94]

A huge range of biocides is available, and those used depend as much on local safety regulations and availability as on recommended efficacy. The four main groups of biocides used in most offshore operations are:

- Chlorine and chlorine-releasing compounds,
- Phenolics,
- Aldehydes, and
- Quaternary ammonium compounds.[95]

Chlorine will kill planktonic bacteria and other organisms at very low concentrations, but up to 2,000 ppm is needed to control bacteria in biofilms as the organic slimes tend to shield the bacteria from its effect.[95] Phenolics and more complex compounds (e.g., pentachlorophenols) possess both bacterial and fungicidal activity. However, many of these, such as chlorinated phenols, are long-term environmental pollutants; thus, they are restricted in use.

Aldehydes, such as formaldehyde and glutaraldehyde, are widely used, though care is needed in handling, and in the conditions in which they are used. For example, while the biocidal activity of glutaraldehyde increases with increasing pH its chemical stability decreases. Bacterial adaptation and possible carcinogenic effects limit the use of formaldehyde, and although glutaraldehyde is more extensively used, it is not an innocuous compound.[94]

Quaternary ammonium compounds (derivatives of ammonium salts, e.g., alkyldimethyl benzyl ammonium chlorides) have varying biocidal activity depending on the architecture of the bacterial cell wall. For example, the amount of one quaternary compound required to be effective varies from 10 ppm for Gram-positive bacteria to 400 ppm for Gram-negative bacteria. These compounds are safe to handle with no known carcinogenic activity, and they are biodegradable once in low (10 to 20 ppm) concentrations. Bessems notes that the known effects of quaternary compounds, and presumably other biocides, on one strain of bacteria are not necessarily the same for another strain.[95] Gaylarde and Johnston have emphasized the enhanced resistance imparted to bacteria when several species are present together on a metal surface at some pH values, and they recommend that mixed bacterial suspensions and metal coupons should be used when assessing biocides.[96]

One major consideration is the need to deal with sessile bacteria and not just those in suspension, as sessile bacteria require greater levels of biocide for control.[95,97] Monitoring is also important in relation to sessile bacteria. Maxwell and Hamilton stress that monitoring the bulk solution does not allow an accurate assessment of the conditions prevalent at metal surfaces, nor are bacterial numbers necessarily a good indication of bacterial activity.[98] They cite a case of the ballast leg of a concrete platform through which several metal structures pass. Studies in the bulk phase showed little or no bacterial activity; however, significant bacterial sulfate reduction was found on metal surfaces indicating potential corrosion problems. Most of the hydrogen sulfide produced by bacteria is formed under sludge or sediment. Therefore, it directly affects the metal surface.

It follows that SRB detected in samples taken from the flowing stream may not show any correlation to free sulfide in the water or to SRB activity and corrosion at the metal surface. The realization that bacterial activity may take place in regions shielded from the moving stream of water by deposits or aerobic bacteria is not new, and in 1963 Wright recommended the assessment of sessile bacteria.[99] The need for careful monitoring of biological activity, together with the use of biocides and corrosion inhibitors is constantly stressed in the literature.[5,8,20,90]

The approach to fouling and corrosion control in water systems usually involves the use of both biocides and corrosion inhibitors. A few seawater lift systems use copper/aluminum biocidal anodes to prevent fouling,[100] but most use chlorine at levels of about 0.5 mg/L.[97,89] However, chlorine itself can cause corrosion problems that reduce pump life to two years.[89] Thus, corrosion-resistant linings or non-metallic pipework are often used for the aerated portions of the system.[90] In filtration systems, a combination of good design and careful selection of the materials from which different parts of the system are constructed and the depth from which the water is taken can produce significant improvements in performance. Bacterial control within the system is by chlorination, up to the deaerators, and thereafter by biocide slugging. Again, increased chlorine dosing has been found ineffective once SRB have become established,[101] and localized SRB activity can only be controlled by use of organic biocide slugs.[97]

Maxwell et al. describe the problems of the control of biological activity in a seawater system in detail.[94] In all systems, good design, careful monitoring and initial chemical cleaning, "pigging" (a "pig" is a device propelled through a pipe to scrape, treat, or monitor the pipewalls), and slug dosing of lines with biocides is essential prior to startup and after any lengthy shutdown to remove scale and debris and start operations with a clean system.

Summary

Although much more investigation into the area is required for a complete understanding, there is little worry in the industry about general corrosion of the external structure under marine fouling, except where this might lead to cracking. The systems used to prevent corrosion (sacrificial anodes and impressed current cathodic protection with or without paint coatings) are generally successful. However, it must be stressed that localized attack, rather than uniform corrosion, is the most normal corrosion morphology encountered in failures, and it is this type of corrosion that is enhanced by biological activity even in the presence of corrosion protection.

It is in the area of hydrodynamic loading and corrosion fatigue that the greatest potential detriment due to fouling exists. Minor structural failures, blamed on the reduction of fatigue life by marine fouling, already have been reported. In the case of older platforms with less design allowance for fouling, the possibility of fouling-induced fatigue increases with the age of the platform. At the very least, this will increase the frequency of inspection needed to ensure safety. Laboratory tests predict rapid failures in hydrogen sulfide environments, but they assume that the crack already exists and is subject to stress. This set of circumstances is more likely to occur as structures get older and corrosion produces pits which grow into cracks.

The effects of biocorrosion on the internal water and oil systems are more apparent and immediate than those on the external surfaces. Careful monitoring, regular maintenance, and the use of suitable treatments can contain the problem, but greater awareness, together with improvements in design and operation, will help the situation.

References

1. A.D. Mercer, A.K. Tiller. R.W. Wilson, eds., *Microbial Corrosion*, book 303 (London, UK: The Metals Society, 1983): pp. 117-122.
2. J.R. Lewis, A.D. Mercer, eds., *Corrosion and Marine Growth on Offshore Structures* (Chichester, UK: Ellis Horwood, 1984): p. 156.
3. S.C. Dexter, *Biologically Induced Corrosion*, NACE-8 (Houston, TX: NACE, 1986).
4. E.C. Hill, J.L. Shennan, R.J. Wilkinson, eds., *Microbial Problems in the Offshore Oil Industry* (London, UK: Institute of Petroleum, 1987).
5. P.F. Sanders, J.D.F. Stott, "Assessment, Monitoring and Control of Microbiological Corrosion Hazards in Offshore Oil Production Systems," CORROSION/87, paper no. 367 (Houston, TX: NACE, 1987).
6. WHOI, "Marine Fouling and Its Prevention" (US Naval Institute, Annapolis, MD: Woods Hole Oceanographic Institution, 1952).
7. J.D. Costlow, R.C. Tipper, *Marine Biodeterioration* (Annapolis, MD: Naval Institute Press, 1984).
8. R.G.J. Edyvean, L.A. Terry, G.B. Picken, "Marine Fouling and Its Effects on Offshore Structures in the North Sea—A Review," *International Biodeterioration* 21(1985): pp. 277-284.
9. R.G.J. Edyvean, "Biodeterioration Problems of North Sea Oil and Gas Production—A Review," *International Biodeterioration* 23(1987): pp. 199-231.
10. K.C. Marshall, "Bacterial Behavior at Solid Surfaces, A Prelude to Microbial Fouling," E.F. C. Somerscales, J.G. Knudsen, eds., *Fouling of Heat Transfer Equipment* (Washington, DC: Hemisphere Publishing Corp., 1981): p. 305.
11. J. Carson, "The Microbiological Aspects of Early Stages of Marine Fouling" (Ph.D. thesis, University of Aston, 1980).
12. M.J. Dempsey, "Marine Bacterial Fouling: A Scanning Electron Microscope Study," *Marine Biology* 61(1981): pp. 305-315.
13. N.I. Hendey, "Littoral Diatoms of Chichester Harbour with Special References to Fouling," *Journal of Royal Microbiology Society* 71(1951): pp. 1-86.
14. R.G.J. Edyvean, B.L. Moss, "Microalgal Communities on Protected Steel Substrata in Seawater," *Estuarine Coastal and Shelf Science* 22(1986): pp. 509-527.
15. L.A. Terry, G.B. Picken, "Algal Fouling in the North Sea," L.V. Evans, K.D. Hoagland, eds., *Algal Biofouling* (Amsterdam, Netherlands: Elsevier, 1986): pp. 179-192.
16. G.N.R. Forteath, G.B. Picken, R. Ralph, "Patterns of Macrofouling on Steel Platforms in the Central and Northern North Sea," J.R. Lewis, A.D. Mercer, eds., *Corrosion and Marine Growth on Offshore Structures* (Chichester, UK: Ellis Horwood, 1984): pp. 10-22.
17. C.S. Johnston, J.C. Side, S.R.H. Davies, "The Use of Environmental Audit in Offshore Operations," E.C. Hill, J.L. Shennan, R.J. Wilkinson, eds., *Microbial Problems in the Offshore Oil Industry* (London, UK: Institute of Petroleum, 1987): pp. 147-163.
18. H.J. Sommerville, "Environmental Aspects of Offshore Operations," E.C. Hill, J.L. Shennan, R.J. Wilkinson, eds., *Microbial Problems in the Offshore Oil Industry* (London, UK: Institute of Petroleum, 1987): pp. 105-124.
19. T.G. Wilkinson, "Biological Mechanisms Leading to Potential Corrosion Problems," J.R. Lewis, A.D. Mercer, eds., *Corrosion and Marine Growth on Offshore Structures* (Chinchester, UK: Ellis Horwood, 1984): pp. 31-37.
20. P.F. Sanders, W.A. Hamilton, "Biological and Corrosion Activities of Sulfate-Reducing Bacteria in Industrial Processing Plant," S.C. Dexter, ed., *Biologically Induced Corrosion*, NACE-8 (Houston, TX: NACE, 1986), pp. 47-68.
21. D.G. Oldfield, "Scale of the (marine fouling) Problem in UK Waters," *Marine Fouling of Offshore Structures* 2 (London, UK: Society of Underwater Technology, 1981).
22. T.L.C. Ingram, "The Maintenance of Offshore Structures," *Offshore* 37(1978): pp. 32-35.
23. J.D. Bultman, C.R. Southwell, C.W. Hummer, "Biocorrosion of Structural Steels in Seawater," *Reviews on Coatings and Corrosion* 2(1977): pp. 187-214.
24. P.F. Sanders, S. Maxwell, "Microfouling, Macrofouling and Corrosion of Metal Test Specimens in Seawater," *Microbial Corrosion*, book 303

(London, UK: The Metals Society, 1983): pp. 74-83.

25. R.G.J. Edyvean, L.A.Terry, "The Effects of Marine Fouling on the Corrosion of Offshore Structures," *Proceedings of the UK Corrosion '84 Conference* (London, UK: Institute of Corrosion Science and Technology, 1984), pp. 195-198.
26. L.A. Terry, R.G.J. Edyvean, "Recent Investigations into the Effects of Algae on Corrosion," L.V. Evans, K.D. Hoagland, eds., *Algal Biofouling* (Amsterdam, Netherlands: Elsevier, 1986): pp. 211-230.
27. R.G.J. Edyvean, "Algal-Bacterial Interactions and Their Effects on Corrosion and Corrosion-Fatigue," C.A.C. Sequeira, A.K. Tiller, eds., *Microbial Corrosion* 1 (London, UK: Elsevier Applied Science, 1988), pp. 40-52.
28. G. Cragnolino, O.H. Tuovinen, "The Role of Sulfate-Reducing and Sulfur-Oxidizing Bacteria in the Localized Corrosion of Iron-Base Alloys—A Review," *International Biodeterioration* 20(1984): pp. 9-27.
29. W.A. Hamilton, "The Sulfate Reducing Bacteria and Anaerobic Corrosion," *Annual Reviews of Microbiology* 39(1986): pp. 195-217.
30. V. Scotto, "Electrochemical Studies of Biocorrosion of Stainless Steel in Seawater," G.J. Licina, ed., Microbial Corrosion, *1988 Workshop Proceedings* (EPRI ER-6345, Research Project 8000-26,) April 1989, pp. 131-1336.
31. S.C. Dexter, "Role of Microfouling Organisms in Marine Corrosion," *Biofouling*, in press 1993.
32. L.A. Terry, R.G.J. Edyvean, "Microalgae and Corrosion," *Botanica Marina* 24(1981): pp. 177-183.
33. A.D. Woolmington, J. Davenport, "pH and pO_2 Levels Beneath Marine Macrofouling Organisms," *Journal of Experimental Marine Biology and Ecology* 66(1983): pp. 113-124.
34. L.A. Terry, R.G.J. Edyvean, "Influences of Microalgae on Corrosion of Structural Steel," J.R. Lewis, A.D. Mercer, eds., *Corrosion and Marine Growth on Offshore Structures* (Chichester, UK: Ellis Horwood, 1984): pp. 38-44.
35. F.G. Hardy, "Fouling on North Sea Platforms," *Botanica Marina* 24(1981): pp. 173-176.
36. S.C. Dexter, G.Y. Gao, "Effects of Seawater Biofilms on Corrosion Potential and Oxygen Reduction of Stainless Steel," *Corrosion* 44, 10(1988): pp. 717-723.
37. A. Mollica, A. Trevis, "Correlation between the Formation of a Primary Film and the Modification of the Cathodic Surface of Stainless Steel in Seawater," *Proceedings of the Fourth International Congress on Marine Corrosion and Fouling* (Juan-Les-Pins, Antibes, France: 1976), pp. 351-365.
38. V. Scotto, R. DiCintio, G. Marcenaro, "The Influence of Marine Aerobic Microbial Film on Stainless Steel Corrosion Behavior," *Corrosion Science* 25, 3(1985): p. 185.
39. S.C. Dexter, H.J. Zhang, "Effect of Biofilms on Corrosion Potential of Stainless Alloys in Estuarine Waters," *Proceedings of the Eleventh International Corrosion Congress* vol. 4(Florence, Italy: 1990), p. 333.
40. R. Johnsen, E. Bardal, "Cathodic Properties of Different Stainless Steels in Natural Seawater," *Corrosion* 41, 5(1985): p. 296.
41. B. Little, R. Ray, P. Wagner, Z. Lewandowski, W. Lee, W. Characklis, F. Mansfeld, "Electrochemical Behavior of Stainless Steel in Natural Seawater," CORROSION/90, paper no. 150 (Houston, TX: NACE, 1990).
42. F. Mansfeld, R. Tsai, H. Shih, B. Little, R. Ray, P. Wagner, *Corrosion Science* 33, (3)1992 p. 445.
43. B. Little, F. Mansfeld, personal communication, 1992.
44. S.C. Dexter, "Implications of Biological Effects to Localized Marine Corrosion of Stainless Alloys," *Proc. First Pan-American Corrosion and Protection Congress*, V. Rascio, ed. (Mar del plata, Argentina: 1992), p. 485.
45. S.C. Dexter, P. Chandrasekaran, H-J. Ahang, S. Wood, "Microbial Corrosion in Marine Environments: Effect of Microfouling Organisms on Corrosion of Passive Metals," *Proc. Second NSF-CONICET Workshop on Biocorrosion and Biofouling* (Memphis, TN: Buckman Laboratories, 1992).
46. A. Mollica, G. Ventura, "Electrochemical Monitoring of the Biofilm Growth and Corrosivity in Seawater: Effect of Intermittent Chlorination," *Proc. 8th Intl. Congress on Marine Corrosion and Fouling* (Taranto, Italy: [in press], 1992).
47. W.A. Hamilton, "The Sulfate Reducing Bacteria: Their Physiology and Consequent Ecology," *Microbial Corrosion*, book 303 (London, UK: The Metals Society, 1983): pp. 1-5.
48. C.O. Obuekwe, D.W.S. Westlake, J.A. Plambeck, F.D. Cook, "Corrosion of Mild Steel in Cultures of Ferric Iron-Reducing Bacterium Isolated from Crude Oil II: Mechanism of Anodic Depolarization," *Corrosion* 37,11(1981): pp. 632-638.
49. R.G.J. Edyvean, L.A. Terry, "Polarization Studies of 50D Steel in Cultures of Marine Algae," *Interna-*

tional Biodeterioration Bulletin 19(1983): pp. 1-11.95.64.

50. N.J. Heaf, "The Effect of Marine Growth on the Performance of Fixed Offshore Platforms in the North Sea," *Marine Fouling of Offshore Platforms* 1 (London, UK: Society of Underwater Technology, May 1981): pp. 19-20.
51. R.W.S.M. Kingsbury, "Marine Fouling of North Sea Installations," *Marine Fouling of Offshore Structures* I (London, UK: Society for Underwater Technology, 1981).
52. J. Wolfram, A. Theophanatos, "The Effects of Marine Fouling on the Fluid Loading of Cylinders: Some Experimental Results," *Proceedings of the Seventeenth Offshore Technology Conference* (Houston, TX: OTC 4954, 1985), pp. 517-526.
53. R. Vaux, "Kelp Adds to Wave Loading," *Offshore Research Focus* 62(1988): p. 12.
54. A. Theophanatos, J. Wolfram, "Fluid Loading on Marine Fouled and Artificially Roughened Cylinders of Various Aspect Ratios," *Proceedings of the Offshore Mechanics and Arctic Engineering (OMAE) Conference* (Houston, TX: OMAE, 1987), p. 47-60.
55. P. Bristol, J.A. Roeleveld, "Fatigue of Offshore Structures: Effect of Seawater on Crack Propagation in Structural Steel," *European Offshore Steels Research Project Seminar, 2* (Cambridge, UK: Welding Institute, 1978), pp. 439-458.
56. U.W. Obuzor, K.R. Gowers, R.A. Cottis, "The Effect of Marine Fouling on Hydrogen Uptake by Metals," C.A.C. Sequeira, A.K. Tiller, eds., *Microbial Corrosion* 1 (London and New York: Elsevier Applied Science, 1988): pp. 53-65.
57. C.J. Thomas, R.G.J. Edyvean, R. Brook, "Biologically Enhanced Corrosion Fatigue," *Biofouling* 1(1988): pp. 65-77.
58. I.M. Austen, E.F. Walker, "Corrosion-Fatigue Crack Propagation in Steel under Simulated Offshore Conditions," ed. C.J. Beever, *Proceedings 2nd International Conference on Fatigue Thresholds, Fatigue '84*, vol. 3 (Warley, UK: EMS Publishers, 1984), pp. 1,457-1,471.
59. T.G. Wilkinson, "Offshore Monitoring," *Microbial Corrosion*, book 303 (London, UK: The Metals Society, 1983), pp. 117-122.
60. L.P. Miller, "Rapid Formation of High Concentrations of Hydrogen Sulfide by Sulfate-Reducing Bacteria," *Boyce Thompson Institute* 15(1949), pp. 437-465.
61. G. Eidsa, E. Risberg, "Sampling for the Investigation of Sulfate Reducing Bacteria and Corrosion on Offshore Structures," S.C. Dexter, ed., *Biologically Induced Corrosion*, NACE-8 (Houston, TX: NACE, 1986), pp. 109-113.
62. R. Ralph, K. Goodman, "Foul Play Beneath the Waves," *New Scientist* 81(1979): pp. 1,018-1,020.
63. K.S. Goodman, R. Ralph, "Fouling—The Marine Growth Industry," *Offshore Engineer* (Sept.,1979): pp. 113-117.
64. J.A. Hardy, J.L. Bown, "The Corrosion of Mild Steel by Biogenic Sulfide Films Exposed to Air," *Corrosion* 40, 12(1984): p. 650.
65. J.B. Davis, *Petroleum Microbiology* (London, UK: Elsevier, 1967).
66. B.A. Crouch, "Biocides in Oilfield Operations," *Biocides in the Oil Industry* (Institute of Petroleum, 1982).
67. B.A. Crouch, "Evaluation of Biocides for Use in North Sea Oil Operations," eds., T.A. Oxley, S. Barry, *Biodeterioration* 5(1983): pp. 725-730.
68. N.S. Battersby, D.J. Stewart, A.P. Sharma, "Microbiological Problems in the Offshore Oil and Gas Industries," *Journal of Applied Bacteriology, Symposium Supplement* (1985): pp. 227s-235s.
69. J.M. Sharpley, "Microbiological Corrosion in Waterfloods," *Corrosion* 17(1961): pp. 380-390.
70. L.L. Wolfson, "Microbiology in Secondary Recovery Systems," *Corrosion* 16(1959): pp. 298-330.
71. G.E. Myers, B.M. Slabyi, "The Microbiological Activity of Injection Waters Used in Alberta Oil Fields," *Producers Monthly* 26, 6(1962): pp. 12-14.
72. U. Steiner, "Effects of Nutrient Accumulation on Heat Exchanger Fouling," *Chemical and Process Engineering* 52(1971): pp. 60-61.
73. A.N. Moosavi, W.A. Hamilton, "Microbial Corrosion Studies in a Marine Sulfuretum," E.C. Hill, J.L. Shennan, R.J. Wilkinson, eds., *Microbial Problems in the Offshore Oil Industry* (London, UK: Institute of Petroleum, 1987): pp.13-26.
74. P. Bos, J.G. Kuenen, "Microbiology of Sulfur-Oxidizing Bacteria," *Microbial Corrosion, book 303*, (London, UK: The Metals Society, 1983), pp. 18-27.
75. D.W.S. Westlake, K.M. Semple, C.O. Obuekwe, "Corrosion by Ferric Iron-Reducing Bacteria Isolated from Oil Production Systems," S.C. Dexter, ed., *Biologically Induced Corrosion*, NACE-8 (Houston, TX: NACE, 1986): pp. 193-200.
76. R.C. Salvarezza, H.A. Videla, "Electrochemical Behaviour of Aluminum in Cladosporium Resinae Cultures," S. Barry, D.R. Houghton, G.C. Llewellyn, C.E. O'Rear, eds., *Biodeterioration* 6 (UK: CAB International, 1986), pp. 212-217.
77. R.W. Mitchell, E.M. Finch, "Water Quality Aspects

of North Sea Injection Water," *Journal of Petroleum Technology* (1981): pp. 1,141-1,152.
78. J.D.A. Miller, R.A. King, "Biodeterioration of Metals," R.J. Gilbert, D.W. Lovelock, eds., *Microbial Aspects of the Deterioration of Materials* (London, UK: Academic Press, 1975): pp. 83-103.
79. R.A. King, R.G. Miller, G.A. Davies, "Corrosion Morphology—Is It Time to Re-Assess Our View of Pitting? " *Engineering Solutions for Corrosion in Oil and Gas Applications* (Milan, Italy: NACE European Conference, 1989).
80. P.D. Gilbert, A.D. Steele, T.D.B. Morgan, B.N. Herbert, "Concrete Corrosion," *Microbial Problems and Corrosion in Oil and Oil Product Storage* (London, UK: Institute of Petroleum, 1983).
81. R.W. Amstutz, L.C. Reynolds, "Water Handling: Key to Flood Success," *Petroleum Engineering* (July, 1961): pp. 82-93.
82. W.B. Bleakley, "Chevron Solves Filtration Problem," *Oil and Gas Journal* (Sept.,1969): pp. 145-148.
83. J.K. Cubine, S.G. Randolf, "Offshore Treating Facilities for Seawater Injection," *Petroleum Engineer* (August, 1973): pp. 38-40.
84. R.G.J. Edyvean, J.A. Pearson, "Fouling of Filters for North Sea Oilfield Injection Water," *International Biodeterioration Bulletin* 18(1982): pp. 117-123.
85. R.R. Matthews, R. Tunaal, P. Mehidizadeh, "Evaluation of Seawater Filtration Systems for North Sea Applications," *Proceedings of the Sixteenth Offshore Technology Conference* (Houston, TX: OTC 4660, 1984): pp. 121-128.
86. R.G.J. Edyvean, A.D. Sneddon, "The Filtration of Plankton from Seawater," *Filtration and Separation* 23, 3(1985): pp. 184-189.
87. B.L. Carlberg, "How to Treat Seawater for Injection Projects," *World Oil* (July, 1979): pp. 67-71.
88. L.C. Case, "Water Problems in Oil Production—An Operator's Manual," 2nd Ed.(Tulsa, OK: PPC Books, 1977).
89. C.C. McCune, "Seawater Injection Experience—An Overview," *Journal of Petroleum Technology* (1982): pp. 2,285-2,270.
90. P.J. Harvey, H.R. Blagden, "Practical Experience of Operating Offshore Seawater Injection Facilities," *Filtration and Separation* 23, 4(1985): pp. 246-249.
91. L.C. Case, "Will Corrosion Eat Up the Waterflood Profit?" *Oil and Gas Journal* (January, 1961): pp. 76-79.
92. B.N. Herbert, "The Effect of Hydrostatic Pressure on Bacteria in Seawater Intended for Injection into Oil Formations," *Journal of Applied Bacteriology* 41(1976): pp. xii.
93. B.N. Herbert, F.D.J. Stott, "The Effects of Pressure and Temperature on Bacteria in Oilfield Water Injection Systems," *Microbial Corrosion*, book 303 (London, UK: The Metals Society, 1983): pp. 7-17.
94. S. Maxwell, K.M. Mclean, J. Kearns, "Biocide Application and Monitoring in a Waterflood System," E.C. Hill, J.L. Shennan, R.J. Wilkinson, eds., *Microbial Problems in the Offshore Industry* (London, UK: Institute of Petroleum, 1987): pp. 209-218.
95. F Bessems, "Biological Aspects of the Assessment of Biocodes," *Microbial Corrosion, book 303* (London, UK: The Metals Society, 1983), pp. 84-89.
96. C.C. Gaylarde, J.M. Johnston, "The Effect of Some Environmental Factors on Biocide Sensitivity in Desulfovibrio: Implications for Biocide Testing," *Microbial Corrosion* (London, UK: The Metals Society, 1983): pp. 91-97.
97. R.W. Mitchell, "The Forties Field Seawater Injection System," *Journal of Petroleum Technology* (1978): pp. 877-884.
98. S. Maxwell, W.A. Hamilton, "The Activity of Sulfate-Reducing Bacteria on Metal Surfaces in an Oilfield Situation," S.C. Dexter, ed., *Biologically Induced Corrosion*, NACE-8 (Houston, TX: NACE, 1986), pp. 284-290.
99. C.C. Wright, "Rating Water Quality and Corrosion Control in Water Floods," *Oil and Gas Journal* (May, 1963): pp. 154-157.
100. "North Seas First Big Flood Project Nears Peak Injection Rate," *Oil and Gas Journal* (August, 1978): pp. 120-128.
101. "Fouling, Corrosion Problem Needs Research," *Ocean Industry* (August, 1978): pp. 67-68.

CHAPTER 8

Identification of MIC

By R.E. Tatnall and D.H. Pope

Is It or Isn't It MIC?

This section addresses the characteristics that distinguish microbiologically influenced corrosion (MIC) from "nonbiological" corrosion. It is important to distinguish MIC from nonbiological corrosion because the solutions to the problem might depend on understanding its causes. For example, if the problem is biological in nature, the corrosion inhibitor being added could be a nutrient for corrosive bacteria so adding more will merely aggravate the problem. On the other hand, raising the operating temperature could slow biological activity dramatically; but if the problem is not biological in nature, to do so would increase corrosion rates.

MIC and other corrosion mechanisms may operate together. It is often difficult to prove conclusively whether or not the biological role is important in a specific case of corrosion. Definite proof usually requires an expensive and lengthy research program, and few people with corrosion problems are willing to fund such research. Most simply want a "fix." For practical reasons, corrosion engineers often have to settle for less than definite proof when making corrective recommendations. The "proof" often comes in observing the response to changes made to try to correct the problem. For example, if increasing the system temperature significantly decreases the problem, a biological role is implied. A good solution at minimum cost requires the best possible understanding of the problem.

Consider the System Chemistry

All MIC reported to date has occurred in the presence of free water or aqueous phase. Even in organic solvent and fuel systems, biological activity is concentrated in water-bearing sludges in tank bottoms or in droplets of water dispersed throughout the system. The corrosion engineer should consider whether or not the system in question contains such free water or if it is an aqueous solution. If neither condition exists, the corrosion problem is probably not biologically influenced. However, the corrosion engineer should consider one caution: Several failures have *initiated* during brief exposure to water, such as during pressure testing of new equipment. Once the *chemical* corrosion cells are thus established, corrosion may proceed to ultimate failure after the equipment is placed in service, even though the normal service chemistry will not support biological activity.

Biological activity, and thus MIC, is strongly influenced by temperature. Most common environmental microorganisms have "optimum" temperatures for growth somewhere in the 20° to 50°C (68° to 122°F) range. Certain other organisms, however, operate most effectively at temperatures outside this range. A few are even active at temperatures of 100°C (212°F) and above. The key point, however, is that any given strain of organism generally has a temperature range of only about 10° to 20°C (50° to 68°F) over which that organism grows and functions best. Above or below that range, that particular organism's activities slow down or stop altogether (but some other organism might flourish at that new temperature). Corrosion electrochemistry, on the other hand, nearly always increases with rising temperature.

These facts provide an opportunity to determine whether or not the corrosion is MIC. If, for example, you know that one end of a heat exchanger is hotter than the other end, and you have a rough idea as to the metal temperatures through the exchanger, look for "patterns" of corrosion related to temperature. Should corrosion be mostly in the band representing, for instance, metal temperatures of 30° to 50°C (86° to 122°F) and decline at higher and lower temperatures, that is evidence that the corrosion is MIC.

If on the other hand the corrosion gets progressively worse over a very broad range, or to temperatures well above 100°C (212°F), it probably is not a case of MIC. The most important key is to look for corrosion severity *decreasing* with increasing temperature, which goes against "normal" electrochemical guidelines.

Observe the Type of Corrosion

With few exceptions, MIC is localized, i.e. pitting or crevice corrosion. However, on bare steel exposed to biologically active water, a general biofouling may occur, under which the MIC "pits" are so close together that they create what is often described as "general, nonuniform corrosion." Even here, the nonuniform "pitted" appearance should arouse suspicion that this might be MIC.

Attempts have been made to relate the surface morphology within corroded areas to the presence or absence of biological factors. However, remember that microorganisms cause corrosion by generating a corrosive chemistry or by helping electron transfer across metal surfaces. Surface morphology relates to the *chemistry* at the metal surface, not to the presence or absence of microorganisms.

For example, a faceted corroded surface on properly heat-treated wrought stainless steel (SS) or the preferential attack of ferrite phase in SS weld metal indicates that corrosion may have been caused by ferric chloride, but not necessarily that such an environment was biologically produced. Is it likely that such a corrosive chemistry was generated by nonbiological factors? If not, *then* microorganisms should be suspected.

Certain patterns of corrosion have become associated with MIC. Rounded pits under discrete mounds (or "tubercles") on carbon steel usually signal MIC, which is appropriately called "tuberculation corrosion." Subsurface cavities along weld seams in SS equipment usually point to iron-oxidizing bacteria. Again, such patterns should arouse suspicions of MIC. However, such patterns alone do not prove anything.

Observe Covering Deposits

Almost all cases of MIC are accompanied by biologically rich deposits over the corrosion site. Assuming these deposits are not cleaned away before observation, they may provide important clues to the cause of the corrosion.

If fresh wet deposits have neutral pH and exhibit a distinct slimy feeling, they probably include bacterial slime ("exopolymer"), indicating that slime-forming bacteria have been active on the surface. Such bacteria are known to harbor corrosive types or may be corrosive themselves. On carbon steel, the biodeposit may contain so much particulate iron oxides and trapped solids from the water that the slimy feeling is completely masked. Many tubercles on steel are hard and brittle, even though they are or were biologically bound.

On metals other than carbon steel, localized, discrete mounds or deposits are indicative of biological activity. However, more evidence than this is needed to prove that the organisms were actually *causing* corrosion. The next logical step in the investigation is to analyze the mound material for elements (i.e. sulfur, phosphorus, iron, manganese, and chlorine) that may be indicative of MIC. The most popular method of analyzing deposits is with a scanning electron microscope equipped for energy dispersive spectroscopy (also known as EDS, EDXA or "EDAX"). One shortcoming of this method is that peaks for sulfur overlap peaks for molybdenum, and the characteristic peak for manganese coincides with the secondary peak for chromium. Where such ambiguity is likely, wavelength dispersive spectroscopy offers the resolution necessary to discern among the various elements.

A word of caution is necessary regarding the use of such x-ray analyses. While they indicate presence of the elements, they cannot be used to determine the form (compound or valence state) of these elements. For instance, a high sulfur peak may indicate a sulfide compound. However, it may just as likely be a sulfate, elemental sulfur, or any other sulfur compound. X-ray diffraction analysis (provided the material is crystalline) or wet chemical analysis may be used to distinguish among these forms. Other more exotic analytical methods are also available. However, it is important to understand the limitations of the methods you use.

On carbon steel or cast iron, tubercles or deposits associated with MIC caused by sulfate-reducing bacteria (SRB) usually contain moderately high levels of phosphorous.[1,2] Sulfur may also be present, but it is usually concentrated in the corrosion product below the covering deposit. This will be discussed later in more detail. As expected, deposits on steel are usually very high in iron.

On alloys other than carbon steel and cast iron, the covering deposits associated with SRB usually contain few distinct characteristics other than high levels of iron and common "water particulates," i.e., silicon and aluminum. If iron oxidizers or manganese bacteria are present, high levels of chlorine, with or without manganese, can be expected. Mounds associated with subsurface pitting along SS welds offer the best example of this. These mounds typically contain very high levels of iron, chlorine, and manganese.[3]

Not all MIC is accompanied by discrete covering mounds. One common phenomenon on SS is subsurface tunneling corrosion under uniform biofilms rich in manganese, especially in chlorinated waters (Figures 9.11-9.14 in Chapter 9, Catalog of Color Photographs.) Sometimes small mounds are seen projecting through the biofilm; in other cases, no mounds are observed.

One final caution about covering deposits or biofilms: These materials consist primarily of inorganic particles trapped from the water or deposited by the organisms. They usually analyze approximately 10 to 15 percent organic by such methods as thermogravimetric analysis, slightly higher by ignition loss in air. When these materials dry out, they become virtually indistinguishable from dried mud. This has caused many biodeposits to be labeled incorrectly as "settled inorganic solids." (How mud can settle out uniformly on vertical or overhead surfaces has not been explained.)

Analyze Corrosion Products

Sometimes, MIC has distinct corrosion products within the corroded areas below the covering biomass. This is especially true in the presence of SRB. These organisms produce sulfide, which is very reactive and tends to combine with any of several metal ions, including manganese, chromium, iron, and copper. To the extent that such metal ions are not available, sulfide reacts with hydrogen to form volatile hydrogen sulfide (H_2S), the "rotten egg" odor usually associated with septic conditions.

Sulfate reducing bacteria, like all bacteria, use phosphorous chemistry to store energy. Therefore, their active regions tend to show significant levels of phosphorous, often in amorphous compounds that defy analysis by x-ray diffraction.[1]

Corrosion products associated with SRB are almost always black if observed fresh from the site. On carbon steel and cast iron, primary elements are iron, phosphorus, and sulfur. On SS, chromium and sulfur are usually prevalent in the corrosion products, while iron and phosphorus may be present in the covering biomass but are usually only trace elements in the corrosion product.[2]

When iron oxidizing bacteria are involved in the corrosion, ferric chloride or ferric-manganic chloride are the most likely corrosive agents. Therefore, reaction products of corrosion are soluble chlorides. Liquid extracted from subsurface cavities in SS may analyze high in iron and chlorine, but analyses of the covering deposits may prove most useful in such cases.

Analyses of both corrosion products and covering deposits often reveal only the base metal(s) in question (e.g. iron, chromium, and nickel in the case of AISI 304 SS) plus chlorine. While these cases may be easy to understand from a corrosion standpoint, they are the hardest to prove regarding biological contribution. Microbiologists have not put forward mechanisms whereby microorganisms might concentrate chlorides. It has traditionally been assumed that these highly mobile ions simply diffuse into deposits and pits in which an electron deficiency exists, as a result of electron-consuming reactions such as chemical reduction.

When chlorides are the only significant anion present in the corrosion product, the corrosion engineer should try again to rationalize their presence based on nonbiological mechanisms. For example, if a deposit is in place and temperatures exceed 100°C (212°F) under the deposit at the metal surface, local boiling might cause concentration of salts in the deposit. However, in such cases, *all* anions in the water, including sulfates and carbonates, should concentrate. If only the chlorides are concentrated, then a biological effect might be suspected. While this concentration is still not *proof* of MIC, finding significant numbers of microorganisms in the covering biomass or at the metal surface will further support the case.

If the deposit or corrosion product contains large amounts of sulfates, it is apparent that sulfur compounds were involved, which leads to obvious questions: Was this equipment handling sulfuric acid? Were sulfuric or sulfamic acids recently used to clean this equipment? If the answers to such questions are no, chances are that biological activity existed. Sulfides may have been present which subsequently converted to sulfates on exposure to air. It is also possible that sulfur- or sulfide-oxidizing bacteria are growing on the metal surface. If this is true, organisms that *produce* that sulfur or sulfide (e.g. SRB) might also be observed.

If an analysis of the corrosion product shows nothing but oxides of the metals involved, it might be assumed that the corrosion is not biological in nature. This is not necessarily true. There are many organisms that generate organic acids, such as acetic, formic, and propionic. The products of corrosion by these organisms will convert to oxides upon contact with air. Even if analyzed fresh, these organic anions are not detectable by many of the common analytical techniques, i.e., x-ray techniques (energy dispersive or wavelength dispersive spectroscopy).

One word of caution is required regarding such organic anions: Many microorganisms routinely produce similar anions. For example, common strains of SRB produce acetate. In environmental situations, another organism (perhaps another SRB) that uses this acetate as its "food" (organic carbon source) will usually be found. It would not be uncommon in such cases to find acetate present in the biomass. This does not necessarily mean acetic *acid* was present. To make the case for organic acid corrosion, free acidity (low pH) must be shown, or organisms known to produce low pH must be identified. This is also true of such organic anions as formate, propionate, succinate, lactate, and butyrate, all of which are known "intermediates" for biological communities.

What about pH?

Removal of the corrosion product from the walls of a sulfuric acid storage tank after it has been rinsed to permit vessel entry will probably indicate no free acidity. This is because iron sulfate is not an acid, and the residual sulfuric acid was soluble and, therefore, rinsed away. The same may be true of MIC. Whether or not free acidity on a microscale existed during the corrosion process, a neutral pH might be measured at the corroded metal surface. For this reason, checking pH during an after-the-fact investigation may be misleading.

There are several MIC situations that are known to produce low pH. One is the case of subsurface cavities in stainless alloys under deposits containing iron-oxidizing bacteria. Using a syringe, ferric chloride solution can be extracted from inside the cavities. The solution will most likely have a pH between 1 and 3. The irony is that iron-oxidizing bacteria apparently have little tolerance for such low pH conditions, which may explain why the covering deposit containing the organisms will usually analyze near-neutral pH.

Other reported low pH cases involve species

of sulfuric acid-producing *Thiobacillus* bacteria. These organisms not only develop free acidity, but some strains grow best in an environment of low pH, from 2 to below 1.

Seeking More Proof of MIC

Imagine that a specific case of corrosion is compared to others presented in this book, and it appears to be a case of MIC. The form of corrosion is then observed, covering deposits and corrosion products are analyzed, and the corrosion still appears to be microbiologically influenced. Samples have been sent to a microbiological laboratory that has confirmed so-called "corrosive" microorganisms. However, skeptics in the investigating team say it is just "water corrosion" or "under-deposit corrosion" and want to treat the system with corrosion inhibitors and dispersants. Others want to make a stronger case for MIC, and thus give biocides, temperature change, or continuous on-line cleaning top priority. Such dilemmas are why so many cases of apparent MIC are "solved" by overkill solutions that satisfy every possible corrosion mechanism.

If the problem is expensive, and the proposed solutions are also expensive, further study may be justified. The best way to further define the problem is to reproduce it in a controlled test, either in the laboratory or in the field. In either case, however, a parallel "control" test must be run to show the true effect of test variables. This is essential in cases of MIC because biological activity is not constant; rather, it may change dramatically with such things as season, rainfall, temperature, and many other factors that are not always apparent.

The key to such a test is to *reproduce* the corrosion problem both in form and in magnitude. This may take weeks or even months to do, which is why gaining support for a scientific test program is often difficult when dealing with people who want an immediate solution to the corrosion problem. If the corrosion engineer succeeds in setting up a meaningful test and reproduces the problem exactly, the first test variable might be temperature. For example, if the system was operating at approximately 35°C (unfortunately, a very common temperature for cooling water systems), a parallel test under conditions unfavorable to MIC, e.g. at 60°C or with antimicrobials added should be run. If the new conditions produce dramatically less corrosion, the corrosion probably included MIC. In addition to such "scientific" studies, a successful test facility is an ideal way to test proposed changes in operating conditions or chemical treatments. (This is discussed more in Chapter 10, MIC Treatment and Prevention.)

Detecting Microorganisms

The previous section addressed analyzing evidence at a corrosion site to obtain proof of MIC. It concluded that identifying microorganisms associated with the corrosion is one of the most convincing ways to substantiate a claim for MIC, as opposed to simple "chemical" corrosion. This section will discuss the need to identify organisms associated with corrosion, how to go about identifying them, and conclusions that can be drawn (and those that *cannot* be drawn) from the results of that analysis. This is not intended to be a training manual on microbiological analytical techniques, as most actual analyses are best left to specialists in that area. However, a few techniques are available for use by nonmicrobiologists, and these will be discussed. The main purpose of this chapter is to advise the corrosion engineer what to expect and what not to expect when having samples analyzed for microorganisms.

Because most MIC problems involve bacteria, this section will focus on that family of microorganisms. However, fungi can also cause serious problems, especially in petroleum fuel systems. Algae are not usually linked directly to corrosion failures, but they may harbor corrosive bacteria. An application of the general principles for analyzing bacteria, outlined here, will probably indicate if other types of microorganisms are present in significant quantities.

Why Analyze for Microorganisms?

Detecting large numbers of microorganisms at a corrosion site helps prove the case for MIC.

However, it is not definitive proof because the organisms might be growing there *because* of the corrosion; that is, they are organisms that derive energy from oxidizing or reducing metal ions produced by the corrosion reaction.

Therefore, in testing for organisms, it is important not only to see high numbers, but also to identify specific types of organisms that reasonably can be linked to the apparent corrosion mechanism. Finding high numbers of SRB *plus* metal sulfide corrosion products certainly makes a strong case for MIC of steels. The presence of *Gallionella* (iron-oxidizing bacteria), subsurface cavities along SS welds, and deposits containing high levels of iron, manganese, and chlorine makes a strong case for MIC of SS.

Therefore, the microbiological analysis must be interpreted *in conjunction with* other evidence found at the corrosion site. However, without confirming the presence of organisms, it may be difficult to convince skeptics that the corrosion is microbiologically influenced.

"Bugs" Grow in Communities, Too

It is unusual to find individual strains of bacteria or fungi growing alone in nature because natural environments generally do not satisfy all of their needs. Therefore, microbes grow in mixed, interdependent "consortia." Understanding this phenomenon is important to understanding the pitfalls of biological analysis and helps to explain why many cases of MIC do not fit "textbook" microbiological scenarios.

Unfortunately, most of what is known about the behavior of microorganisms has been learned by studying them in pure cultures in the laboratory, under conditions far different than those in the real world. Many microbes grow and function under a wide variety of conditions, responding differently to differing conditions. One example of this is the common sulfate-reducing genus *Desulfovibrio*, which is usually associated with the reduction of sulfate to sulfide accompanying the oxidation of organic nutrients, all of which gives these organisms energy and matter required for growth. In this "normal" respiratory metabolism, they use various short-chain fatty acids and produce acetate. However, if deprived of sulfate, some *Desulfovibrio* strains may become fermenters and produce hydrogen and carbon dioxide instead. By understanding such behavior, it can be accepted that sulfate reducers may be active in a system free of sulfate or sulfides and may, in fact, be supporting the activity of other bacteria, such as methane-producers. This also explains why SRB might be active and plentiful in a system that shows no evidence of sulfide production.

Testing Water Samples vs Testing Surface Deposits

When gathering evidence about a corrosion failure or evaluating the corrosion likelihood in a system, the corrosion engineer may have access to samples of biomass attached to surfaces or may have to settle for samples of the water or process fluid. Both sample types have advantages and disadvantages and it is important to know the limitations of each.

Deposits from the corrosion site are the first choice for testing purposes because the offending organisms are expected to be found in relatively high numbers at the corrosion site. However, surface deposits are not homogeneous; rather, they usually consist of layers or zones containing different types of organisms and chemical species. This is especially true in aerated waters, where the outer portion of the deposit usually has a relatively high population of aerobic slime-forming bacteria, such as *Pseudomonas*. Deeper in the center of the mass, higher numbers of slime-forming bacteria may be found. It would not be surprising to find a few filamentous iron-oxidizing bacteria scattered throughout the upper and middle layers, especially if the biomass is covering a corrosion site on an iron-containing alloy. SRB and other anaerobic bacteria, such as methanogens or *Clostridia*, will be most abundant in the interior portions of the biomass.

Because of the potential for such varied findings, the best sampling technique for deposits in aerated systems might be taking separate samples from the outer surface, the middle mass, and the area nearest the metal surface. If analyzing that many samples is not practical, care should be taken, at least, that all "zones" are represented in the sample.

The second choice for obtaining testing samples is water or process fluid from the system in question. Such samples have one major drawback: The numbers and types of organisms present in the fluid may have little correlation to those on the metal surfaces. If a system is in equilibrium, with no changes occurring in temperature, organic nutrient levels, and flow patterns, and if the system is not being treated with biocide, a correlation might exist between the number of free-floating (planktonic) organisms and the number of sessile organisms growing on surfaces (sessile). If "stressful" changes are occurring in the system (temperature leaving the optimum range for growth, nutrients diminishing, etc.), growth on surfaces is favored, and the relative number of planktonic organisms will diminish. Most biocides are more effective in killing planktonic organisms than sessile organisms; therefore, samples of systems treated with biocides may indicate low numbers in water samples but high numbers growing on surfaces.

With these limitations in mind, the corrosion engineer can see that testing water samples is a compromise, but it is sometimes the only option, and tests of water samples are preferable to not monitoring a system at all.

The third sampling strategy is to test samples of biofilms taken at random from surfaces. This approach has experienced increased popularity with the proliferation of devices designed to monitor biofouling tendencies by periodic removal of sample plugs from a sampler wall. While this technology is useful in monitoring general *biofouling* tendencies, it is probably of limited usefulness in monitoring relative MIC tendencies since localized MIC "cells" may occur randomly on surfaces and may form regardless of the tendencies to form biofilm in the system.

How Microorganisms are Detected

Chemists, metallurgists, and materials scientists are accustomed to analyzing metals, chemicals, or corrosion products, and they expect to learn exactly what elements and compounds are present. While such findings may require a combination of methods, it is possible to discover, with a degree of certainty, the chemical makeup of a material. However, identification of organisms is not as exact a science. Classical methods, still used, are analogous to the wet chemical methods once used predominantly for chemical analyses. Only now are new and direct methods for identification and enumeration of organisms becoming available for routine use. It is important that the corrosion engineer knows what methods are presently available, and to understand their limitations.

Culture Methods

Most microbiological analyses attempt to identify organisms by lumping them into a few general categories. Almost all classical methods of identification involve "culturing" the sample by placing it in or on a chemical nutrient medium, incubating it for a period of time, and observing for signs of microbial growth.

If testing for SRB, for example, the microbiologist would typically use an oxygen-free medium containing lactate and sulfate. The use of such a medium presumes that only sulfate reducers will thrive in this particular combination. Therefore, if colonies grow the microbiologist assumes the colonies are sulfate reducers. To be sure that other anaerobic, lactate-utilizing bacteria are not being counted erroneously, some reduced iron might be added to the medium. Then, any black precipitate that forms is assumed to be iron sulfide, overwhelming evidence that sulfate was being reduced and, thus, that sulfate reducers were active. However, significant color could take several weeks to develop. Until about 1950, though, such testing was state of the art; therefore, it was accepted. Microbiologists have recently learned that some of the more common SRB cannot use lactate, but they can use acetate. Additionally, these strains grow very slowly in the laboratory, so that detection using classical culturing techniques could take months. Fortunately, newer media are being developed that propose to detect these hard-to-grow strains in less time.

Culture methods for most aerobic bacteria are easier and faster than those for SRB, but they also have drawbacks. The most common method of culturing for aerobic organisms uses a semi-solid agar medium (agar is a clear, gelatin-like substance derived from seaweed) containing nutrients of almost universal appeal to microorganisms. The growth surface is left in contact with air, a condition under which it is assumed that obligate anaerobes will not grow. After a few days of incubation, the number of colonies is counted. The result is related to the original sample volume to estimate the total number of aerobic organisms per unit volume in the sample. In making this estimation, it is assumed the organisms were thoroughly dispersed and that colonies did not result from initial clusters of bacteria. This is rarely the case when using real-world samples.

The color and texture of the various colonies is noted to estimate the numbers of organisms of each type. Some types also can be determined by detecting characteristic odors from their colonies. Because of such "sophisticated" techniques, for example, most slime samples from cooling water systems are diagnosed as containing high numbers of *Pseudomonas*. This diagnosis is of little practical use because, rather than being a single strain, *Pseudomonas* is a diverse genus of common, hardy aerobic organisms associated with everything from industrial biofouling to urinary tract infections.

Another risk in identifying organisms by culture techniques stems from the conditions used for culture, which sometimes differ from actual conditions of the environment from which the sample was collected. Taking an unknown sample to determine what will grow on a specific medium, at a specific temperature, within a specific time, creates the possibility that certain types of organisms in the sample will grow to large numbers while others will not grow at all.

The final outcome of this analysis may bear little resemblance to the population that existed in the sample's original environment. Identifying *Pseudomonas* bacteria, for example, is of little use if they only make up one percent of the original population while the organisms that were originally predominant, which were actually causing the problems, simply may not thrive in those particular culture conditions.

With this risk in mind, the corrosion engineer should consider the limitations of using culture results *or any other single technique* to make a diagnosis of MIC.

Direct Microscopic Methods

Culture methods have traditionally been used for two reasons: Most bacteria are very small and require a microscope to be seen, and their shapes are not particularly distinctive. In environmental samples, the rationale for using culture methods is reinforced by the presence of large quantities of solid "debris" which make many microscopic analyses difficult. However, because of disadvantages cited in connection with the culture techniques, some types of bacteria are best detected using direct microscopic techniques.

Most practical manuals on microbiological analysis of environmental samples stress that *all* samples should be observed under a microscope. To observe organisms among the large amount of debris usually found in deposit samples, the sample is placed on a glass slide so that microbes stick to it and stained to make bacteria distinguishable.

Various staining techniques can be used, depending upon the sort of organisms to be observed and the type of illumination provided by the microscope. Some stains distinguish only certain classes of organisms, while others will stain all bacteria. The latter stain type includes fluorescein isothiocyanate (FITC), which stains all bacteria and glows under epifluorescent illumination. This stain type is used by some laboratories to estimate total numbers of bacteria in samples.

Direct microscopic inspection is also useful in qualitatively determining the presence of filamentous iron bacteria, stalked iron oxidizers (such as *Gallionella*), and certain other organisms of distinctive shape. However, most types of bacteria appear either as spheres (cocci) or rods (bacilli), and little can be determined about them by direct microscopic observation.

In summary, direct microscopic inspection can be useful in estimating total numbers of bacteria by using staining/counting techniques, and direct microscopic inspection is the preferred means of identifying certain larger, distinctively shaped organisms. However, direct microscopic inspection will not identify the vast majority of environmental bacteria and, therefore, should not be used alone in microbiological analysis.

Adenosine Triphosphate Assay

All living cells contain a substance called adenosine triphosphate (ATP), which is used to store energy for cell growth. Several years ago, a method of testing environmental samples for the presence and amount of ATP, and thus estimating the total number of microorganisms in the sample, was developed. The method involves filtering or otherwise removing extraneous matter from the sample, adding a chemical to extract the ATP from any cells present, adding a second chemical (extracted from firefly tails) that reacts with ATP to produce photoluminescence, and reading the light derived with a sensitive instrument called a photometer.

The ATP assay method has some drawbacks. For example, it cannot distinguish between ATP extracted from bacteria and that from leaf particles or other organic debris in the sample. Also, the amount of ATP per bacteria cell, while fairly predictable in some microorganisms, is not predictable in SRB and some other common environmental bacteria. This means that the "total bacteria numbers" derived from this method are likely to be very rough approximations.

Despite these limitations, the method has been used with apparent success to estimate relative total bacteria numbers in oil field water samples, samples in which nonbacteria ATP is rare. ATP assay does offer certain advantages over culture methods. Because it is a direct test, it avoids the concerns related to growing organisms "in captivity." It also offers results in a very short time—less than an hour if samples are tested in a field laboratory near the sampling site.

Specific Antibody Tests

The immune systems of mammals generate antibodies specific to most foreign substances entering the animal's bloodstream. These antibodies will attach only to the particular structure (epitope) against which they are produced. The job of these antibodies is to "label" the foreign substances so that white corpuscles can find them and destroy them before the foreign substances can proliferate. This elegant defense mechanism, essential to the survival of animals, also provides powerful tools in identifying specific types of microorganisms. Two basic types of specific antibody "probes" exist: those using antibodies against the surfaces of whole cells, and those directed at unique proteins within the cells.

Used for many tests in the medical area, specific antibody tests for organisms associated with MIC are currently offered commercially only for the detection of SRB. A cell-surface antibody test, using an epifluorescent microscope to interpret results, is available for use in a laboratory by trained personnel. Relative SRB numbers are determined with direct counting FITC staining methods described earlier, except that in this case, only SRB fluoresces in the microscope field.[6] This method uses cell-surface antibodies, so only SRB strains for which antibodies are present are detected. As for any method, it is important to check the applicability of this technique in a given system by comparing results obtained against other methods. Once it is determined that the test detects the types of SRB present in that system, it may be used with confidence to track relative numbers with time. It is rapid (<1 hour) and inexpensive (once the microscope is purchased).

Another SRB test kit is available that tests not for SRB cells per se but rather for the presence of APS reductase, an enzyme present in all SRB, regardless of strain. This test is suitable for use in the field by relatively unskilled personnel, and it requires only 20 minutes for results. The method develops a color endpoint, with the degree of color proportional to the numbers of SRB in the sample.[7]

The sensitivity limit of these antibody methods is *normally* approximately 10^3 cells/mL in the sample. While this figure may sound high to people who have been using culture techniques to detect lower numbers, experience with these methods indicates that they detect SRB in many cases where the "more sensitive" culturing methods indicate much lower numbers. For example, in systems in which acetate-utilizing *Desulfobacter* predominate, the antibody methods report much higher total numbers than do conventional culturing techniques.

One negative aspect of these antibody methods is that they cannot distinguish between living and dead bacteria. This may not be a serious drawback in environmental samples, however, because dead bacteria usually don't hang around long in nature, but are broken down to provide "food" for other organisms.

Fatty Acid Analyses

Most families of bacteria have distinctive fatty acid (lipid) compositions, and at least two types of identification systems are available to identify and semiquantitate organisms by lipid "fingerprinting." The first system is a commercial instrument package sold with computer software and a library of characteristic lipid profiles. This identification method is offered by various commercial analytical services because few individual user companies can justify the cost of the system.

The second fatty acid method specifically analyzes the phospholipids associated with cell membranes. This identification system has been used in research programs for several years, particularly in identifying SRB, which have very unique phospholipid "signatures." While the commercial system mentioned above has somewhat limited sensitivity, the University of Tennessee Oak Ridge National Laboratory has assembled a system that performs this analysis with extreme sensitivity.[8]

The fatty-acid analytical methods have one distinct advantage over antibody techniques: They are nonspecific in that a pattern will be obtained, almost regardless of the type(s) of organisms present. This means that one is less likely to "miss" types of organisms for which one is not specifically testing.

On the negative side, the fatty acid methods are relatively expensive. They tend to be less quantitative than the methods mentioned earlier, indicating "high numbers," "low numbers," or "none detected" rather than the order-of-magnitude numbers determined by other methods. Additionally, the current library of fatty-acid signatures of environmental organisms is limited. This means that many peaks in the output curve are likely to be detected, but it may not be possible to associate them with any known organisms. Some types of bacteria associated with MIC, such as the iron oxidizers, do not have unique membrane-bound phospholipids and, thus, are not detected by that particular signature technique.

DNA Probes

Every species of microbe has a unique base sequence (code) in its deoxyribonucleic acid (DNA). Because this coding is constant, regardless of where or how that organism has been growing, DNA analysis could be a powerful tool for identifying specific types of organisms. Although much research is currently being conducted to perfect DNA sequencing techniques for clinical and environmental samples, no practical tests applicable to MIC are commercially available at this time.

Taking and Preserving Samples for Biological Analysis

The methods used to preserve and ship biological samples depend on the analytical techniques to be used. Obviously, it is always best to use fresh, wet samples immediately upon removal from the operating system. Samples obtained in any other way are a compromise because some species may rapidly die off or grow, thus changing the population profile. Nonetheless, the only samples available may be ones that have been sitting around for days or weeks, or even dried deposit materials that are several months old.

These materials may be worth testing, keeping in mind that such samples most likely will not represent the entire population and should not be used as sole basis for diagnosis.

Sampling for Culture Methods

If organisms must grow to be identified, there is one requirement: Maintain original "field" conditions if possible; otherwise, suspend metabolism and preserve viability (the *ability* to grow when conditions are right). Samples taken from cooling water or waste treatment systems should be put in a plastic bag or bottle. For solid samples, add a little water from the actual sampling site. If possible, hold the sample at "normal" operating temperature and get the sample "into culture" (inject it into serum bottles or apply it to agar medium) within hours. If the sample is left sitting in the container too long, the organisms may die.

If the sample cannot be placed into culture within hours (if, for instance, the sample is to be shipped to a remote laboratory for analysis), it should be packed in ice and shipped by the fastest method. The sample should not be frozen because freezing will kill many of the organisms. If the sample came from an aerated system, some airspace should be left in the container so it will not become oxygen-depleted.

The problem with shipping *deposit* samples with water and free airspace is that the deposit material may become dispersed in the water during shipment. A preferred method for deposit samples is to place the deposit material, with a *small* amount of system water, in a thin polyethylene bag (e.g., sterile "whirlpack"), squeeze out most of the air before sealing the bag, and place the bag inside a larger jar or box inside the shipping container. Any loose packing material may be used to restrain the bag. Plastic foam "peanuts" are ideal for this. Enough oxygen will permeate the plastic film to supply the needs of aerobes in the sample for a day or so.

Remember that all procedures for shipping samples for culture tests are compromises. The questions raised and the complexities of the shipping process have prompted many sites to perform culture on the spot. Once the bottles have been inoculated or the agar plates streaked, shipping conditions are less important, provided the samples are protected from freezing and exposure to temperatures higher than the temperature from which they were collected.

Sampling for Microscopic Inspection

Because microscope techniques do not require continuing growth of the organisms, it is necessary only to preserve the cells physically intact. For optical microscopy, preserving with buffered formaldehyde is usually recommended, whereas glutaraldehyde is most often recommended if electron microscopy is to be used. Approximately two percent final concentration of either works well.

Once the samples are so preserved, they should be stable for a few days at room temperature. For long-term storage, refrigeration is recommended, even with preservative added. This preservation technique also applies to the cell surface antibody/epifluorescence test for SRB, except here formaldehyde is preferred.

Sampling for Other Detection Methods

If the sample will be used for the direct enzyme antibody test for sulfate reducers, the test will normally be run within an hour or so after taking the sample. Preservation or special handling is therefore not an issue. For longer storage, the sample should be iced or refrigerated (*not* frozen) until the test is run. The rule of thumb with this method is to test within 4 hours of taking the sample, or within 24 hours if the sample is iced.

Sample preservation for the ATP assay is similar in that any action that will affect the level of ATP in the sample should be avoided. Removing the sample from its natural environment will most likely stress the bacteria in the sample. Many of these organisms react to such stress by suspending their normal metabolism and drawing off their ATP reserves instead. Refrigeration of the sample, as was recommended for the enzyme antibody method, may minimize this effect.

The fact that the ATP assay measures total ATP in the sample—including that which is in the bulk sample fluid—offers another alternative. A chemical that draws the ATP out of the cells into solution can be added, and the sample can be boiled to kill other enzymes that would normally destroy the released ATP. Samples so prepared need not be iced.

When taking samples for fatty acid analysis (including phospholipids), it is not important to keep cells intact, only to prevent cell growth during handling and to avoid things that might destroy or mask the fatty acids present. Freeze-drying is one means of accomplishing this and has the advantage of not requiring special treatment during shipping. If equipment for rapid freeze-drying is not available, the samples can simply be frozen and then placed in a vacuum desiccator to dry. There may be acceptable chemical preservatives, depending on the specific analytical methods planned, but these should be discussed with the persons who are to conduct the final analysis.

Final Comments on Microbiological Analyses

While painting a fairly pessimistic picture of analyzing environmental samples for microbiological makeup, this section has presented the combined experience of consultants who have had to deal with real-world MIC problems. The corrosion engineer should take heart in the knowledge that new methods are being developed all the time, and some of the limitations mentioned here may already have been overcome. In the meantime, results of microbiological analyses should be used with other chemical and metallurgical data to diagnose MIC. Questions regarding methods and interpretation should be asked of anyone who proposes to have an absolute knowledge of what organisms were present at a MIC site and in what relative numbers.

This discussion should offer an appreciation that no one sampling and preservation technique is good for all samples, sites, or types of analyses. Before taking important samples, consider what analytical methods might be applied and, if necessary, preserve multiple samples in different ways so as to keep the various analysis options open.

Finally, this section should raise questions about the relationships of specific types of bacteria to corrosion. Many papers, proposing a study of MIC using pure strains in laboratory tests, have ended with comments that the corrosion rates observed in the laboratory rarely match the rates observed in the real world. The MIC community now generally accepts that severe MIC cases are caused by complex communities of organisms, working together, rather than a single pure strain. Therefore, multiple methods should be used to detect as many types of microbes as reasonably possible.

References

1. P.J. Weimer, et al., "Effect of Phosphate on the Corrosion of Carbon Steel and the Composition of Corrosion Products in Two-Stage Continuous Cultures of *Desulfovibrio desulfuricans*," *Applied and Environmental Microbiology*, (1988): pp. 386-396.
2. R.E. Tatnall, "Experiments in Microbiologically Influenced Corrosion." CORROSION/84, paper no. 95 (Houston, TX: NACE, 1984).
3. G. Kobrin, "Corrosion by Microbiological Organisms in Natural Waters," *Materials Performance* 15 (1976): pp. 38-46.
4. E.S. Littmann, "Oilfield Bactericide Parameters as Measured by ATP Analysis," *International Symposium of Oil Field Chemistry* (Society of Petroleum Engineers of AIME), paper no. 5312, (Dallas, TX: AIME, 1975).
5. E.S. Littmann Jr., "Use of Adenosine Triphosphate to Monitor Microbial Quality of Hydroxypropyl Guar Based Fracturing Fluids," Nalco Chemical Co. Report.
6. D.H. Pope, "Discussion of Methods for the Detection of Microorganisms Involved in Microbiologically Influenced Corrosion," ed. S.C. Dexter, *Biologically Induced Corrosion*, NACE-8 (Houston, TX: NACE, 1985), pp. 275-283.
7. R.E. Tatnall, K.M. Stanton, R.C. Ebersole, "Methods of Testing for the Presence of Sulfate-Reducing Bacteria." CORROSION/88, paper no. 88 (Houston, TX: NACE, 1988), p. 15.

8. N.J. Dowling, F. Widdel, D.C. White, *Phospholipid Ester-Linked Fatty Acid Biomakers of Acetate-Oxidizing Sulfate-Reducers, and Other Sulphide-Forming Bacteria* (1985).

9. R. Prasad, "Pros and Cons of ATP Measurement in Oil Field Waters." CORROSION/88, paper no. 87 (Houston, TX: NACE, 1987), p. 14.

Guidelines for Destructive and Nondestructive Examination of Potential MIC-Related Failures

By S.W. Borenstein

Introduction

Establishing the exact cause of corrosion is an important step in alleviating the problem. Microbiologically influenced corrosion (MIC) is recognized as a major cause of pitting corrosion failures in natural waters. If there is a possibility a specific failure is a result of MIC, a careful failure analysis must be conducted because MIC can be misdiagnosed as attack resulting from conventional chloride crevice/pitting corrosion unless specialized techniques are used.

Such techniques include a careful visual examination, *in situ* bacterial sampling of residual water, bacterial and chemical analysis of deposits and/or corrosion products, and radiographic and metallographic examinations. Other techniques, including culture growth, scanning electron microscopy (SEM), and ultrasonic examinations, may be helpful. Metallographic examination can reveal MIC characteristics such as selective pitting of one phase in duplex stainless steel (SS) weldments.

The methods of failure analysis and their importance will be addressed in this section, including various steps in nondestructive examination and destructive analysis.

Nondestructive Testing

MIC has several visual characteristics, including these for austenitic SS:

- Pitting in low points of tanks or piping, where a likelihood of remaining stagnant water exists;
- Concentric rings of rustlike stain or deposit around each pit (in many cases, reddish-brown mounds over pits); and
- MIC pits in SS are often at, or adjacent to, welds with minute pinholes at the surface opening to large subsurface cavities.

Radiographs show variations in film density and are sensitive to defects as small as 2 percent of the wall thickness. Therefore, a pit would be indicated on the film. However, because of difficulty with interpretation, radiography is generally not suitable for measuring pit depth but is extremely useful for detecting subsurface cavities often associated with MIC. Note, for example, Figures 9.4 and 9.5 (page 91).

Nondestructive testing may be used to determine the extent of corrosion. For example, consider a case of through-wall pitting in a SS piping weld. Radiographs indicating large subsurface cavities may be made of nearby welds. Radiographing the other welds can rapidly identify problem areas requiring further investigation before through-wall pitting occurs.

Eddy-current inspection is used to examine nonmagnetic alloy condenser tubing for pitting, cracking, and wall thinning. A coil conducting alternating current is inserted in the condenser tube. The coil sets up eddy currents in the tube by electromagnetic induction, and instrumentation detects and records changes in the eddy currents caused by pits and other discontinuities. Reference standards, containing flaws of known size and shape, are used for comparison with indications from actual flaws to aid in determining approximate depth of pits, wall loss, and so forth.

Destructive Examination

After sampling water and corrosion products, documenting with color photographs, and completing preliminary examinations and nondestructive testing, it is often necessary to remove the failed portion for further examination. Records and sketches should be made to show the locations of the cuts. Use of a log sheet is recommended to identify orientation of the failed section with respect to:

- Direction of flow,
- Top surface,
- Identification numbers,
- Observations (colors, odors, etc.), and
- Last date of service, date of removal, etc.

A brief description of service conditions also should be included.

Before cutting, the surfaces of the failed section should be protected and kept dry. Cut well away from the affected area to avoid altering the metal structure and avoid using a coolant other than clean water. Radiograph the as-received sections. One applicable examination method is the standard given in ASME Boiler and Pressure Vessel Code, Section V, Article 2. Use permanent marker ink or a vibroengraving tool to mark sections where radiographic lead numbers are located. These markings may be used to help locate pits for cross-sectioning.

Both the inside and outside surfaces of the removed section should be photographed. Depending on appearance and type of attack, higher magnifications may be warranted. Remove any deposits with a light brushing, and photograph again. If the deposits are tightly adherent, light glass bead blasting (10 to 30 psi) may be used.

To illustrate examination methods, consider a failure by pitting in an austenitic SS pipe weld. Mark the sections for cutting (Figure 8.1), using the radiograph to locate the pits. Make a tracing of the pipe and mark the tracing in a similar manner. Number each pit and provide additional information (location of discoloration, whether or not this is a through-wall pit, where the top of the pipe is,

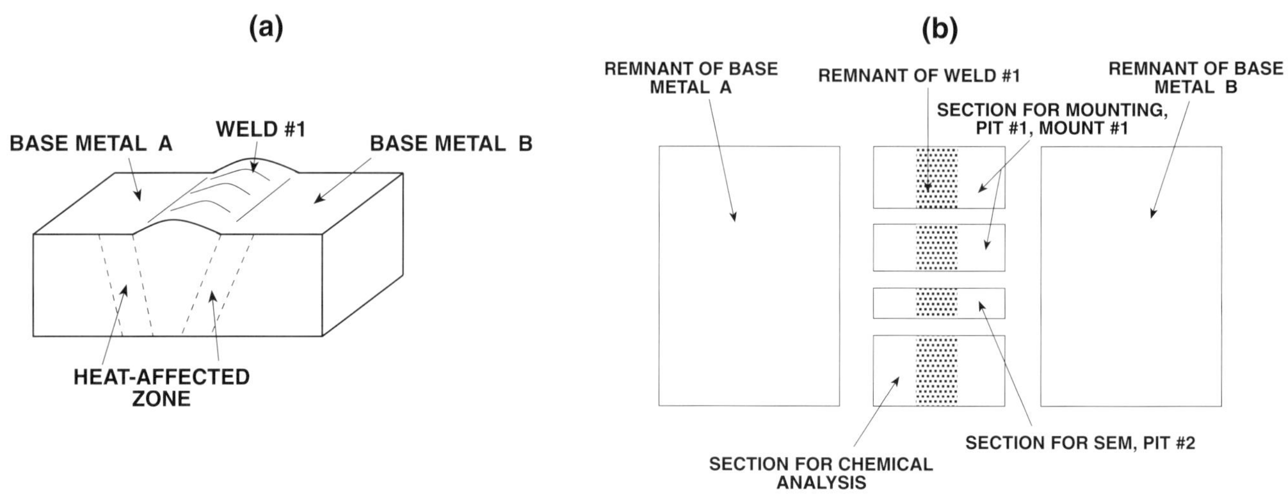

FIGURE 8.1
Sectioning and identification for analysis (a) cross-section of weld (b) sectioned metal.

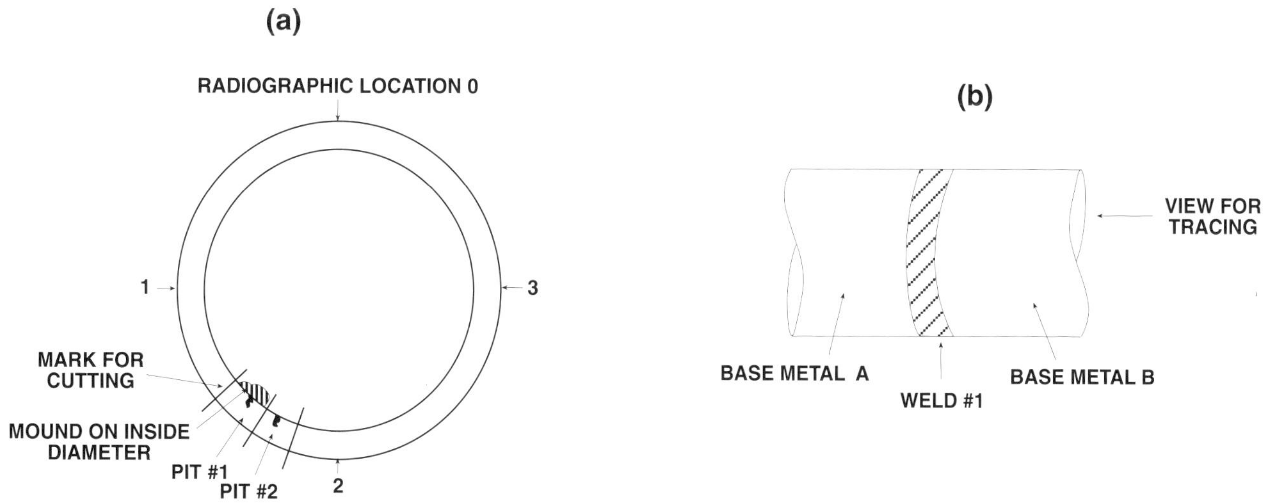

FIGURE 8.2
Tracing information from radiograph (a) end view (b) side view.

direction of flow, etc.). Identify where sections are to be cut for mounting, chemical analysis, and SEM. Cut to one side of a pit (close to, but not into, the pit unless it is quite large). Assign a mount number, and identify each pit on the tracing.

Cut sections for metallography, SEM, and chemical analysis (Figure 8.2), keeping all the pieces attached by a thin section of metal. Photograph the cut sections, including remnants. Identify remnants and all the sections so the original orientation is retained, i.e., which is base metal A and which is base metal B, and so forth.

Prepare metallographic mounts. After polishing, but before etching, examine mounts using optical microscopy. Take photomicrographs at approximately 5X and 100X to examine for selective phase attack before etching. Etch and continue with metallographic examination, taking photomicrographs at 5X, 100X, and 500X. Record information about the attack, i.e., whether or not the attack is in the weld, the heat-affected zone (HAZ), the fusion line, base metal A, and so forth.

Examine the exposed pit with the SEM. Many cases of MIC of austenitic SS weldments display a selective attack of one phase of the duplex structure. Energy dispersive x-ray analysis (EDXA) can be used to identify contaminants (such as chlorides) in the pit.

Contact prints of the radiographs of the pitted regions should be made. The prints should be marked as precisely as possible where the cuts were made, at the identification numbers, and at the mount numbers of the cross-sectioned pits.

COLOR PLATES - Section I

MIC in the Pulp and Paper Industry

FIGURE 4.1
View of the deposits consisting of biomass and debris found in paper machine white water. The deposits (slime) are adhering to the surfaces of the fourdrinier section of the machine. Underneath the deposits are extensive areas of corrosion, both pitting and lateral, on mild steel and SS.

FIGURE 4.2
Surface view of a forged steel flange cover with a residue of solids from the white water and corrosion byproducts. The corrosion as well as the deposition are a result of the growth of slime-forming bacteria on the surface of the flange cover.

FIGURE 4.4
View of the tile-lined stock chest after the fiber/filler was removed. Note that the sulfide-containing substances remain on the surface of the tile.

FIGURE 4.3
Right: View of the surface of a tile-lined stock chest covered with a layer of fiber/filler. Note the deposition of a black material consisting of sulfide-containing substances. The sulfides are produced by anaerobic bacteria growing underneath the layer of pulp.

FIGURE 4.5
View of a carbon steel pipe section covered with corrosion by-products and tubercles formed by the growth of an iron-oxidizing bacterium, *Gallionella sp.* This pipe was part of the treated fresh-water make-up system and had been in service only 18 months.

FIGURE 4.6
View of a carbon steel pipe section from a paper machine white water system. Note the tuberculation and corrosion sites at the low-flow areas.

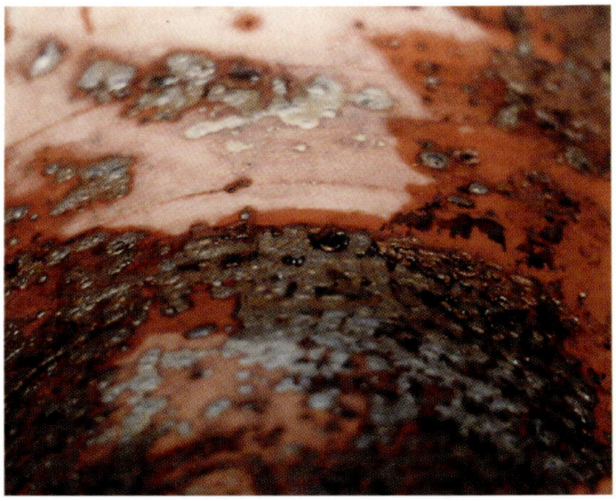

FIGURE 4.8
View of the internal surface of a SS white water pipe after cleaning. Note the extensive pitting and localized deposition of sulfide-containing materials (black deposits). Prior to cleaning, the pit surfaces were covered with microbiological deposits (slime) from which sulfate-reducing bacteria were isolated. The significance of the "red" color is not known. There were several through-wall penetration pits in this pipe segment.

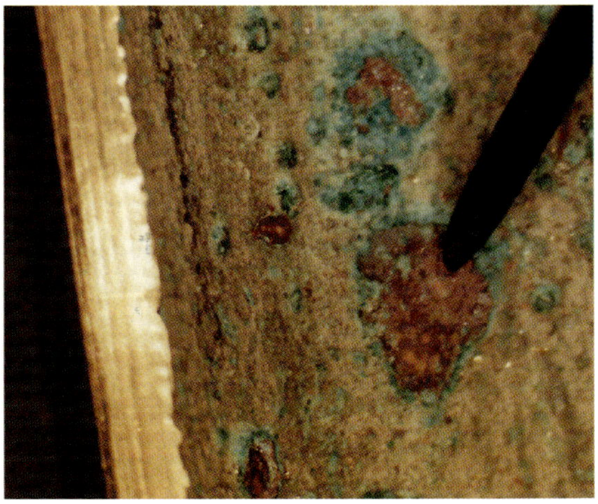

FIGURE 4.7
View of a sectioned admiralty brass tube taken from a freshwater economizer unit. Note the severe pitting, the colonization of growing slime-forming bacteria (at the pointer), the deposition of copper salts at the pit borders, and the general deposition of corrosion byproducts on the tube surface.

FIGURE 4.9
View of a SS couch roll shell failure due to metallurgical stress caused by MIC. The bacteria were growing on the inner surfaces of the suction holes. (Figure 4.10)

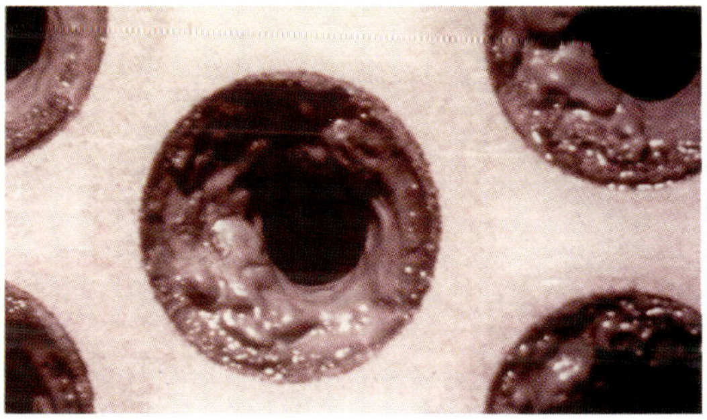

FIGURE 4.10
Close-up view of the biomass deposited on the inner surfaces of the suction holes on a SS couch roll shell. (Figure 4.9)

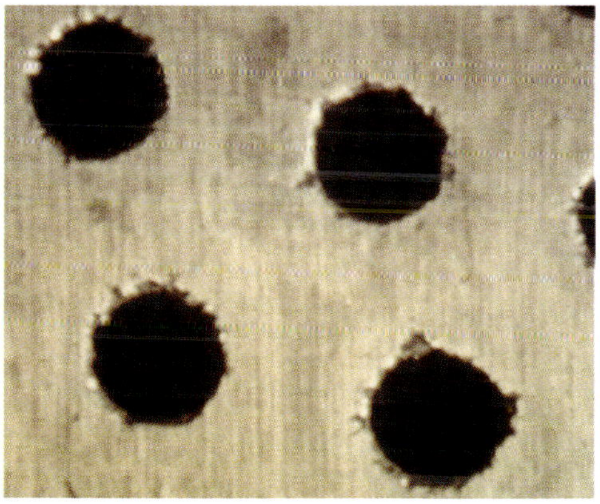

FIGURE 4.11
Close-up view of the outer surface of a SS couch roll on which MIC has occurred. Note the metal loss and deterioration at the borders of the suction holes.

MIC in the Metalworking Processes and Hydraulic Systems

FIGURE 5.1
Soluble oil emulsion split into oil and water phase as a result of bacterial growth. Notice typical tan color of water phase.

FIGURE 5.2
Progressive blackening of soluble oil emulsions from bottom of container indicating growth of SRB.

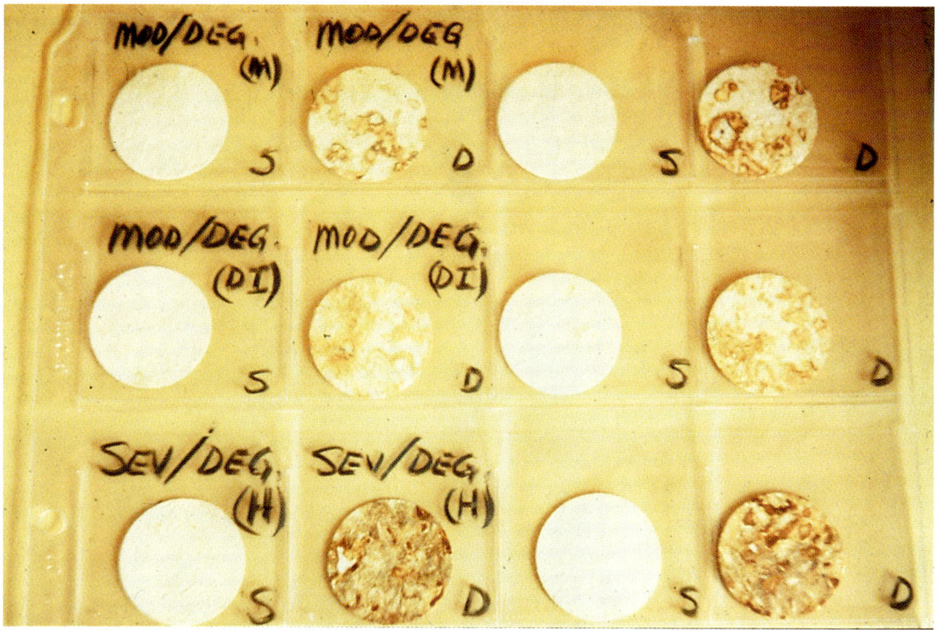

FIGURE 5.4
Cast iron chip corrosion test in soluble oil emulsion. Mod/Deg = aerobic count approx. 5 x 10^7 cfu/mL; SRB approximately 10^3/mL; emulsion visibly intact. Sev/Deg = aerobic count >10^8 cfu/mL; SRB approximately 10^5/mL; emulsion visibly split. M = moderate water hardness (225 ppm Ca++); H = heavy water hardness (450 ppm Ca++); DI = deionized water; S = chips submerged overnight; D = chips decanted after five minutes.

FIGURE 5.3
Control of fungal growth (*Cladorsporium resinae*) in a seawater/hydraulic mixture. No inoculum added. Incubation was three months at room temperature. Top container is control. Mass at interface is *C. resinae*. Middle container has 0.1% morpholine. Notice fuzziness at interface. This is *C. resinae*. Bottom container has 0.2% morpholine. No detectable growth.

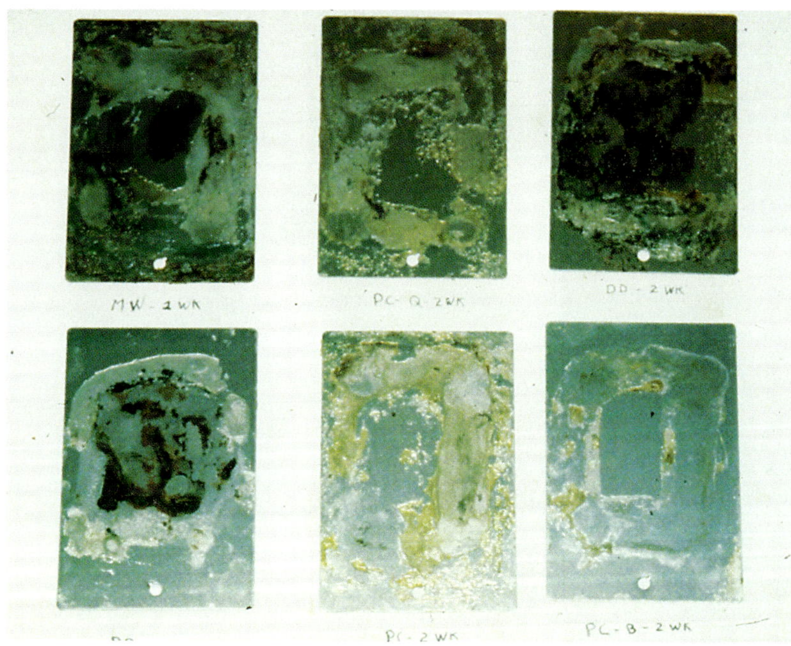

FIGURE 5.5
Control of black staining of 1020 cold-rolled steel after two weeks in vaspar sandwich. Upper left: SRB culture + rolling oil emulsion; upper center: SRB + $NaNO_2$ solution + coating oil; upper right: SRB culture; lower left: SRB culture + coating oil; lower center: SRB culture + $NaNO_2$ solution; lower right: SRB culture + $NaNO_2$ solution + biocide.

FIGURE 5.6
Biofouled, fluted steel filter from a hydraulic press containing invert emulsion fluid.

FIGURE 5.7
Fluted steel filter from a hydraulic press showing rust corrosion after removal of biofilm.

FIGURE 5.8[a]
Simulated filter in invert emulsion inoculated with contaminated hydraulic fluid. After two weeks, notice the extensive fouling of the filter.[8]

FIGURE 5.8[b]
Simulated filter in invert emulsion inoculated with contaminated hydraulic fluid but treated with biocide.[42]

Catalog of Color Photographs of MIC

Compiled by G. Kobrin

This section contains color photographs of conditions—typical biofilm and biomound deposits and corrosion morphologies—that relate to observed cases of MIC. Also included are a series of photomicrographs of algae, fungi, and bacteria known to be associated with MIC.

Most of these photos have not appeared in the technical literature but are from the personal files of the following:

1. G. Kobrin
2. R.E. Tatnall
3. S.W. Borenstein
4. R.W.Lutey

They are referenced by the above numbers following each photo caption.

This section is included in the manual to help the reader identify and analyze potential MIC problems. Many additional color photographs, most taken from the literature, appear in the Case Histories section.

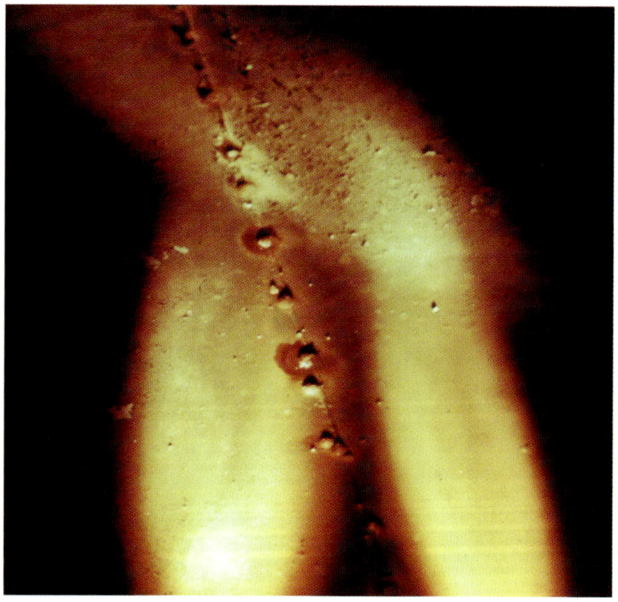

FIGURE 9.1
Biomound deposits along weld seam in bottom of 304 (UNS S30400) stainless steel (SS) tank after several months' exposure to stagnant, ambient temperature well water used for hydrostatic testing and ballast. Deposits were rich in iron- and manganese-utilizing bacteria (*Gallionella, Siderocapsa*), and iron and manganese compounds, predominantly chlorides.[1]

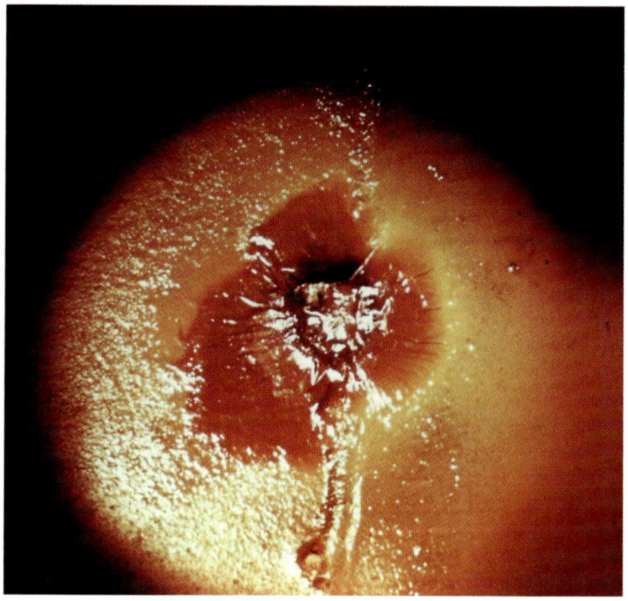

FIGURE 9.2
Closeup of Figure 9.1. Approximate size of deposit is 8 x 10 cm (3 x 4 in.).[1]

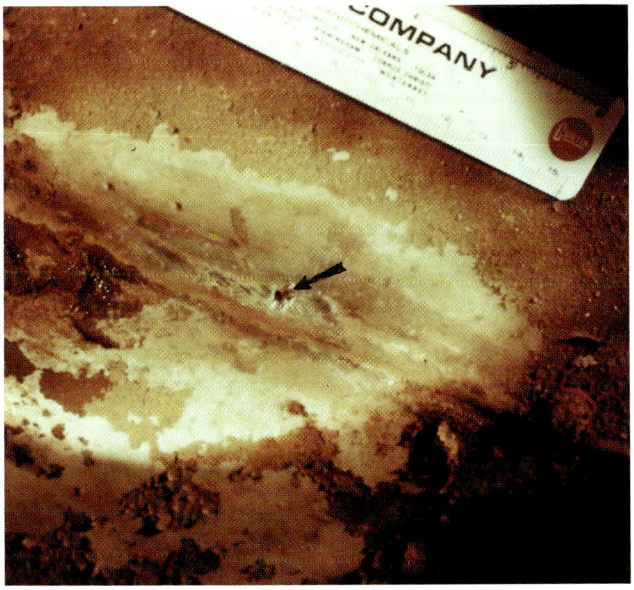

FIGURE 9.3
Pit (arrow) at edge of weld after removal of deposit shown in Figure 9.2 and probing area with an ice pick.[1]

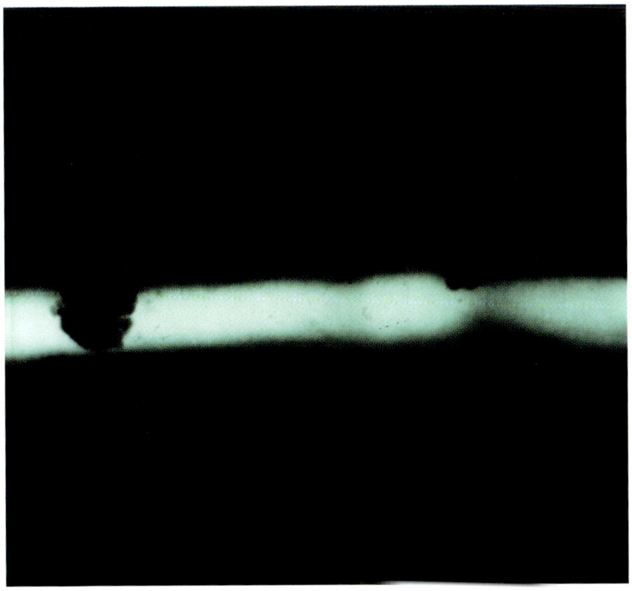

FIGURE 9.4
Radiograph of weld seam shown in Figure 9.3. Note attack is confined primarily to weld metal.[1]

FIGURE 9.5
Cross-section through a pit in a weld seam under one of the biomound deposits shown in Figure 9.1. Note the minute opening (arrow) on the surface as the entrance to a large roundish subsurface cavity beneath a thin shell of metal, typical of MIC at 300 series SS welds.[1]

FIGURE 9.6
Curious rust-colored vertical streaks along internal circumferential weld seams in a 316L (UNS S31603) SS tank exposed for several weeks to the same conditions described in Figure 9.1.[1]

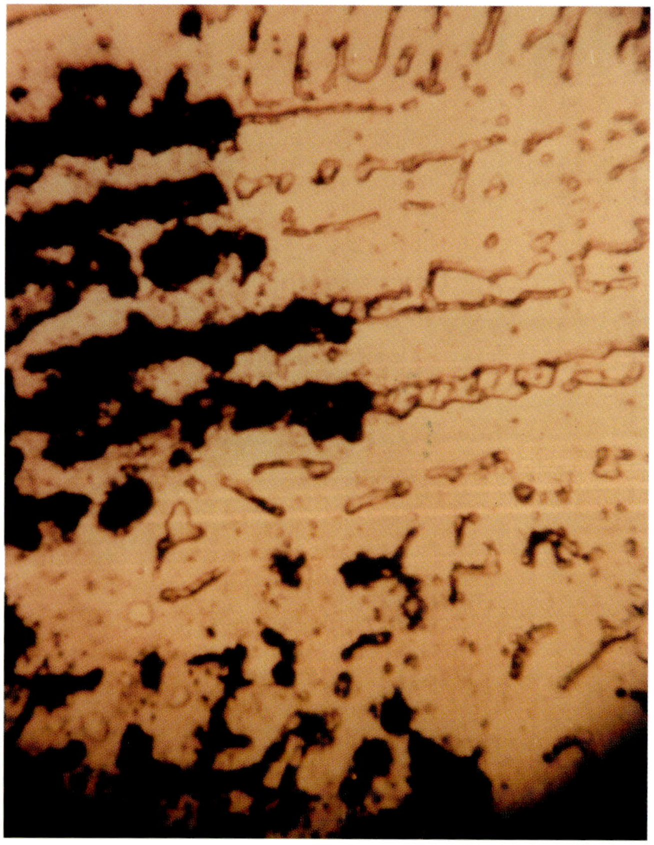

FIGURE 9.7
Closeup of streaks shown in Figure 9.6. Note biodeposits and pits at weld associated with streaks.[1]

FIGURE 9.8
Left: Photomicrograph of cross-section through a pitted 316L SS weld seam shown in Figure 9.7, indicating preferential attack of delta ferrite stringers in the weld, another characteristic of MIC at 300 series SS welds. Approximately 200X. Electrolytically etched in 10% oxalic acid.[1]

FIGURE 9.9
External streaks emanating from a 304 SS weld under internal iron bacteria (*Gallionella*) biomound deposits in a recirculating cooling water system using river water as makeup.[2]

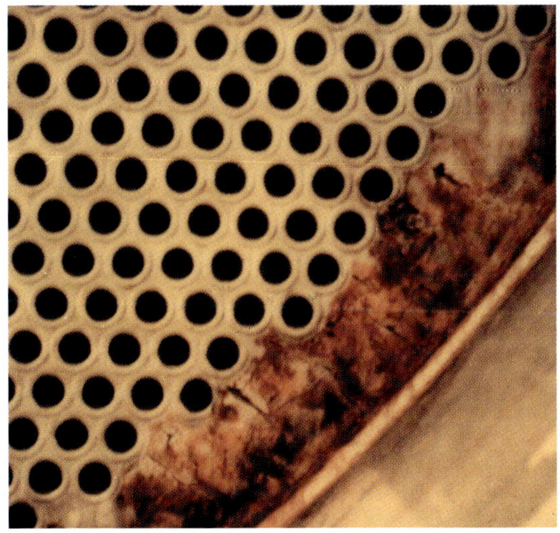

FIGURE 9.10
Surface of 308 (UNS S30800) SS weld-clad tube sheet in a condenser after removal of heavy biomound deposits (unidentified iron/manganese-utilizing bacteria). Weld cladding was severely corroded after approximately 5 years' exposure to cooling tower water with raw river water as makeup.[1]

FIGURE 9.11
Internal surfaces of several 3/4 in. O.D. 304 SS tubes from a condenser exposed for one year to cooling tower water with raw bayou (fresh) water as makeup. Rust-colored stains that remained after high pressure water cleaning are believed to be sites of biomound deposits from unidentified iron/manganese-utilizing bacteria.[1]

FIGURE 9.12
Closeup of one of the sites shown in Figure 9.11. Note pits associated with site. Approximately 3X.[1]

FIGURE 9.13
Radiograph of two of the MIC sites shown in Figure 9.11. Minute openings (pits) on the internal surfaces were entrances to tunnel-like cavities in the 0.083 in. tube walls, one of the characteristics of MIC of 300 series SS parent metal. Approximately 5/6 actual size.

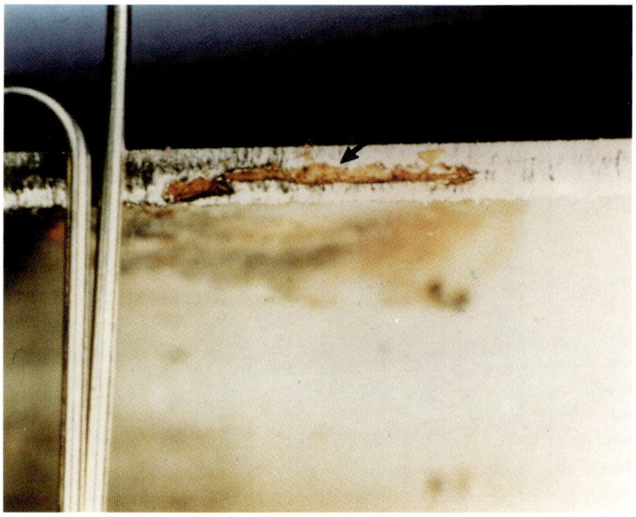

FIGURE 9.14
Longitudinal cross-section through one of the tunnels (arrow) shown in Figure 9.13. External surface is along top edge of tube. Paper clip at left edge shown for size comparison. Approximately 3.5X.[1]

FIGURE 9.15
Wet, slimy biomound deposits (unidentified bacteria) on plain carbon steel tube sheet in heat exchanger exposed to once-through cooling water. Tubes are 90-10 cupro-nickel.[3]

FIGURE 9.16
Broad, shallow craterlike pits under deposits shown in Figure 9.15.[3]

FIGURE 9.17
Pit and perforation in internal surface of plain carbon steel pipeline exposed to sweet crude oil for eleven months. Pit morphology is typical of corrosion by sulfate-reducing bacteria (SRB). Approximately 1.5X.[4]

FIGURE 9.18
Microbiological slime being backwashed from a heat exchanger. Note biomass and black slime deposits containing sulfides adhering to plain carbon steel tube sheet.[4]

FIGURE 9.19
Microbiological slime caused by the uncontrolled growth of both fungi and bacteria in a cooling water system.[4]

FIGURE 9.20
Cooling water intake screen partially fouled with biomass and other debris.[4]

FIGURE 9.23
Right: Closeup of carbon steel surface under fresh tubercles after chemical cleaning. Note bright metal surfaces, indicative of active corrosion.[4]

FIGURE 9.24
Underside of a tubercle removed from a corroded carbon steel surface. Note cavity in center and black material identified as iron sulfide lining the cavity, plus ring- or layerlike structure of tubercle.[4]

FIGURE 9.21
Plain carbon steel waterbox and tube sheet from a heat exchanger exposed to raw fresh cooling water. Bright orange biomounds were produced by the iron bacterium *Gallionella*. Darker biomounds are mature tubercles also produced by *Gallionella*. Black material is iron sulfide produced by SRB in the anaerobic conditions beneath the deposits.[4]

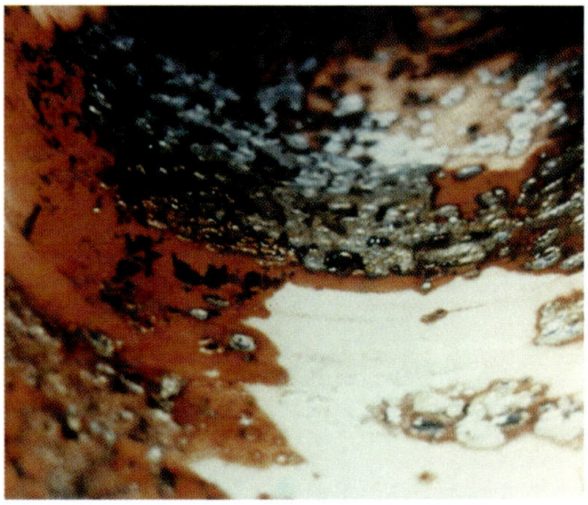

FIGURE 9.22
Left: Closeup of internal surface of 304 SS piping after chemical cleaning. Pitting and perforation were caused by SRB activity. Black material associated with pits is iron sulfide.[4]

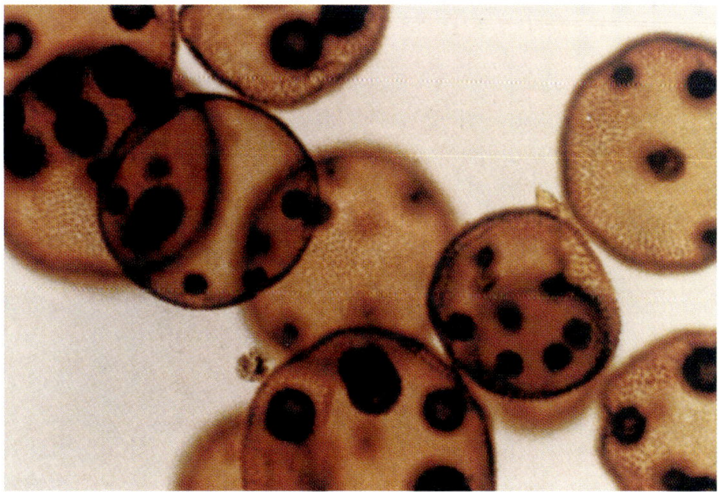

FIGURE 9.26
Photomicrograph of green algae *Volvox*, unstained, at 400X original mag. This algae is relatively large and provides bulk to algae biomass deposits.[4]

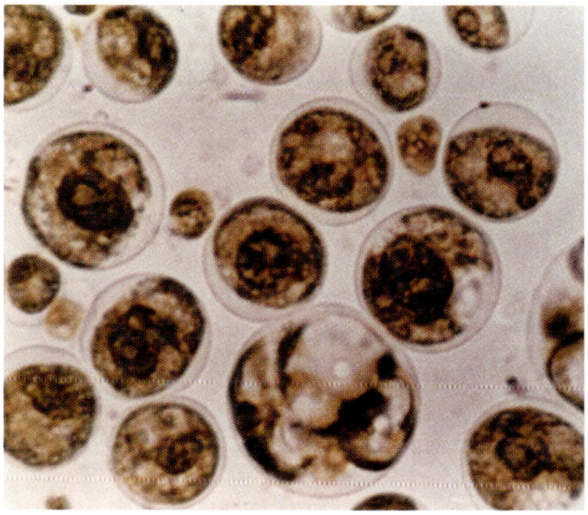

FIGURE 9.25
Photomicrograph of unicellular green algae *Chlorella*, unstained, at 1000X original mag. Note slime layers outside cell walls.[4]

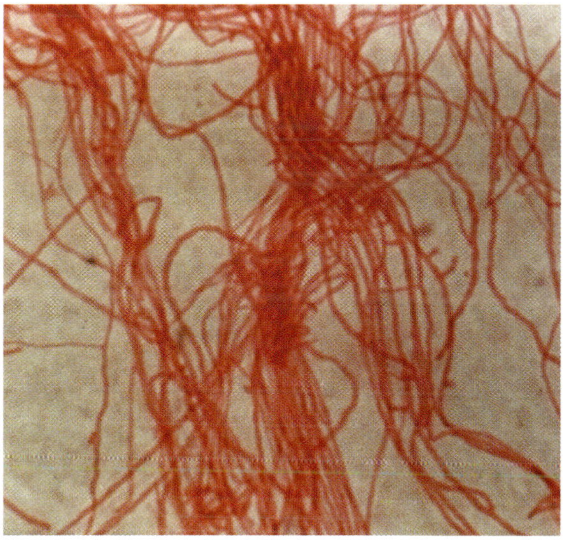

FIGURE 9.28
Photomicrograph of a fungal colony, stained, at 400X original mag. Note filamentous and branching characteristics.[4]

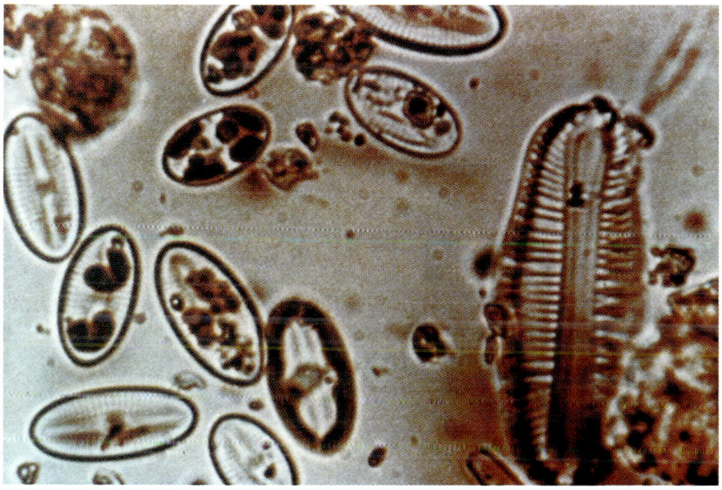

FIGURE 9.27
Photomicrograph of a mixed population of diatoms at 1500X original mag. Note variety of shapes and cell wall configurations.[4]

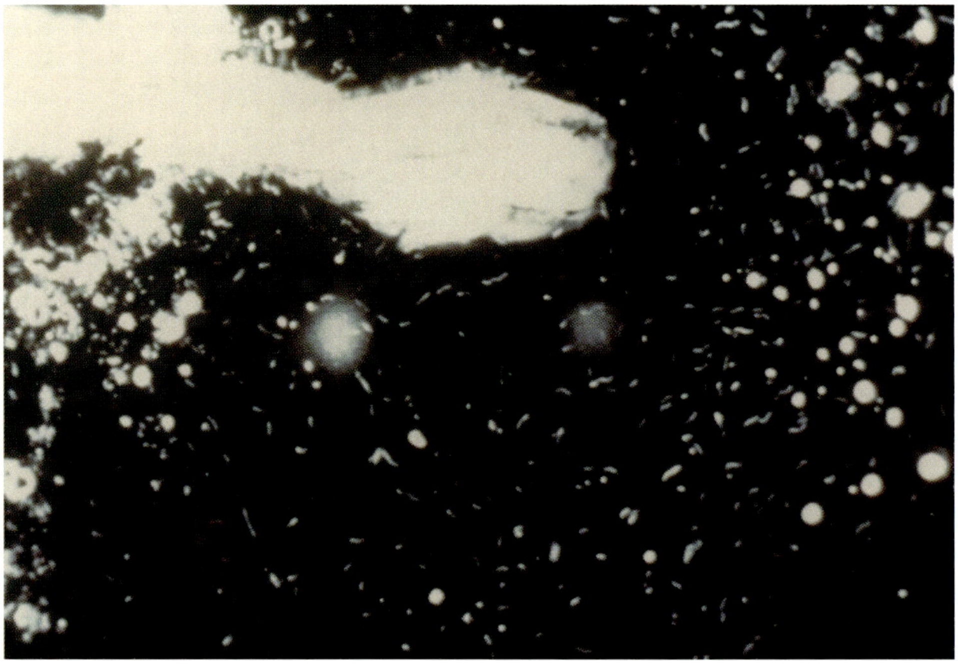

FIGURE 9.30
Dark field photomicrograph of black fluid taken from a pit on carbon steel. Note spirillum cells identified as *Desulfovibrio sp.* (SRB). 1000X original mag.[4]

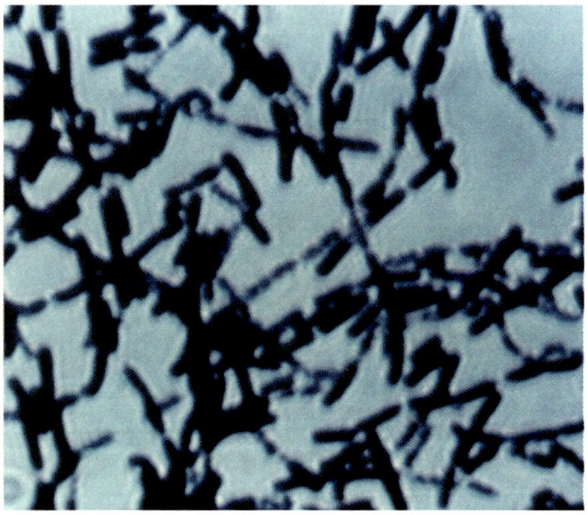

FIGURE 9.29
Photomicrograph of large rod-shaped bacteria, stained, at 1000X original mag. Note tendency to form chains of cells. Blurred appearance is indicative of encapsulation in a layer of extracellular slime.[4]

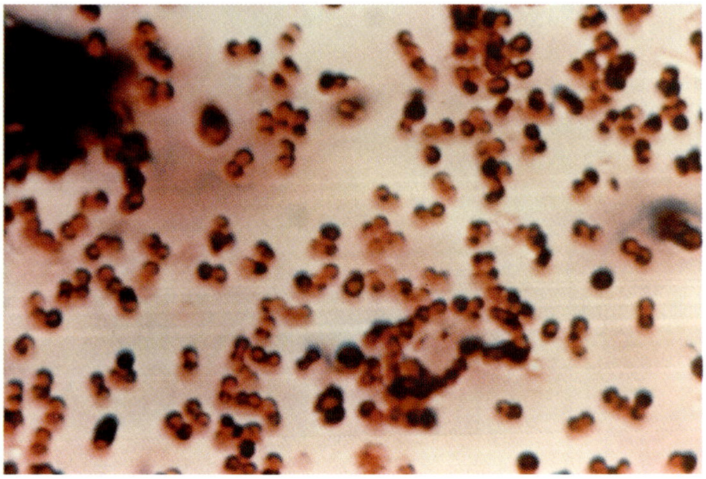

FIGURE 9.31
Photomicrograph of a suspension of SRB *Desulfovibrio sp.*, dark-field illumination, at 1000X original mag. Note curved cells and cell mass adhering to cotton fiber (large white object). Taken from a pit where active anaerobic MIC was occurring.[4]

FIGURE 9.32
Photomicrograph of SRB, *Desulfovibrio sp.*, stained, at 1000X original mag., taken from culture. Note spiral shaped and curved rod-shaped cells and variation in size as they occur in culture.[4]

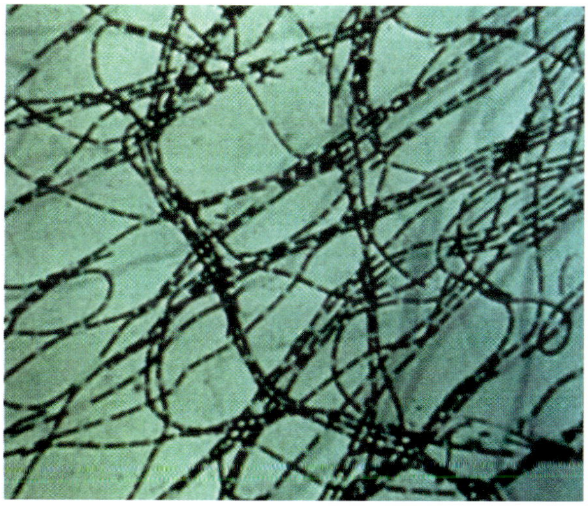

FIGURE 9.33
Photomicrograph of filamentous iron-oxidizing bacteria, *Sphaerotilus natans*, stained, at 1000X original mag. Note strandlike colonization by twisting into a form that can entrain and hold dirt and other debris, resulting in deposits that foul surfaces and create corrosion cells.[4]

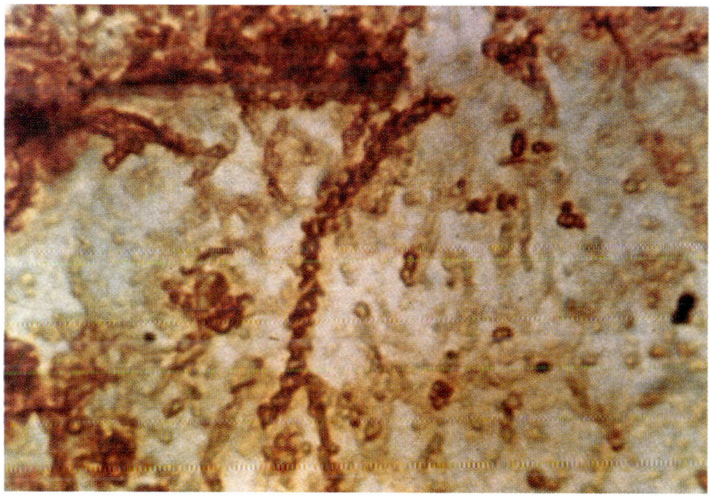

FIGURE 9.34
Photomicrograph of twisted strands of iron hydroxide produced as a byproduct of the iron-oxidizing bacteria *Gallionella sp.*, not stained, at 1500X original mag. Tubercles from these strands cause fouling and MIC.[4]

FIGURE 9.35
Internal surface of admiralty brass condenser tube exposed to once-through river water, showing through-wall pit (arrow) associated with a *Siderocapsa sp.* biofilm. Approximately 4.5X.[4]

FIGURE 9.36
Closeup of surface in Figure 9.35 showing dried biofilm lifting off of tube surface, exposing greenish-black corrosion products of copper oxide and copper chloride.[4]

MIC Treatment and Prevention

By A.A. Stein

Introduction

Microbiologically influenced corrosion (MIC) affects many structural materials in a variety of environments and systems. MIC has occurred in wetted oil storage tanks and oil transmission lines, on the soil side of buried equipment, and in stagnant fire mains, heat exchanger components, cooling towers, cutting-oil emulsion handling, and waste-disposal operations.

Most problems are due to localized corrosion resulting in pitting attack. MIC often occurs in areas where conventional corrosion processes are thought to be under control. These types of corrosion and resultant material degradation have been studied by various industries for years. Recently, it has been recognized that corrosion caused by microorganisms can occur even in very clean systems.

Effective MIC prevention requires identification of the bacteria and verification that the bacteria are a factor in corrosion, even though other corrosion mechanisms cannot be excluded. An extensive corrosion investigation should be performed to assure that the more common corrosion processes are not also occurring. MIC should be considered whenever extensive, repetitive pitting corrosion continues, even with an "upgraded" material. This section presents the various methods of treatment, detection, enumeration, and characterization of the associated bacteria and methods to prevent MIC.

Service Experience

A system's first encounter with conditions that may lead to MIC usually occurs during its initial exposure to an aqueous environment, such as during hydrostatic testing or wet lay-up. Once-through open systems have a different microbiological condition than recirculating systems. The open system will tend to establish a more natural biology, usually with a wider range of organisms than the closed system, where a more limited variety of organisms tend to predominate.

MIC's effects in soils can be predicted based on moisture, pH, and the growth of bacteria. The criteria for the likelihood of corrosion due to bacteria, in terms of total bacteria counts per gram of soil, are presented in Table 10.1.

Table 10.1
General Criteria For MIC in Soil (After Ronay et al.[9])

Bacteria	Bacteria Counts Per Gram of Soil (10^3) Severe	Moderate	No MIC
Aerobic sulfur	15	10 -15	10
Anaerobic (sulfur)	13	8 -13	8
Iron	6	6	3
Sulfate reductants	10	5 -10	5

Detection of MIC

The detection and confirmation of MIC is not straightforward. Judgment is required in the selection of the test methods and in the interpretation of the test data. For example, analyzing the bacteria population in the bulk fluid does not provide a good indication of microbial corrosion. High bacterial counts may not indicate high levels of corrosion because the bacteria may not have attached themselves to the metal surface. Also, low bacterial counts in the bulk fluid do not reflect the levels of bacteria that may exist under deposits. In most cases, microbiological analysis is required to determine the extent of a microbial infection and the consequent potential corrosion hazard.

Regions of particular susceptibility are:

- Dead legs,
- Flow regions less than 5 ft/s continuous flow, or stagnant,
- Areas of pH < 10.5,
- Temperature < 210°F (98.9°C),
- Areas where organic materials are present (debris and so forth) which can provide shelter and nutrients for microbes, and
- Systems and components with interrupted operation (lay-up, maintenance, shutdown).

The occurrence of MIC is usually, but not always, indicated by the presence of nodules. Systems can be inspected for the presence of nodules or corrosion caused by MIC visually or via nondestructive testing (NDT) techniques such as radiographic and ultrasonic inspection. Visual inspection for nodules is particularly useful for stainless steel (SS) systems because MIC usually occurs at welds, and the presence of nodules should be easily observed in an otherwise relatively clean system. Experience has shown that when nodules occur on SS they are most likely bacterial in origin. NDT methods are required to quantify the damage. The relevance of damage to continued plant operation requires use of fracture mechanics and/or stress analyses.

In carbon steel systems, visual inspection alone is inadequate because tuberculation normally present on pipe walls can be caused by either MIC, general corrosion, or both. Nondestructive examination (NDE) methods such as ultrasonic inspection and radiography must be used to determine the extent and severity of corrosion. The inspection frequency is difficult to determine because the initiation and pit growth rates are not predictable.

Flow measurement cannot be used as a method to detect MIC activity. However, high continuous flow velocities (> 15 ft/s) during normal operation is a valid criterion for excluding a system or portions of a system from concern about MIC damage. Flow measurement may be required to assure that system components are receiving adequate flow.

Bacterial and Corrosion Monitoring[6]

All systems should be monitored during installation, modification, and maintenance for the presence of MIC-causing bacteria. Monitoring should include water and biofilm sampling to determine residual biocide, bacterial counts, and biofilm development on surfaces. The bacterial content can be monitored using direct detection techniques. These direct methods provide rapid results because they do not require incubation of a culture. The direct methods include techniques that use adenosine triphosphate (ATP), reductase antibodies, and epifluorescence/cell surface antibodies. The latter method has been used in the field. New methods are continually developed for field application. The common types of bacteria, growth requirements, and corrosion concerns are presented in Table 10.2.

Coupons do not always reveal MIC because they project into the bulk water, and the flow conditions around the coupons may prevent settlement and adherence of bacteria. If pitting attack is indicated, particularly at the edges of samples with no obvious biological growths, the corrosion may not be MIC.

An effective corrosion monitoring practice is the placement of test coupons in the process stream or in a test sidestream. These coupons should be removed for periodic analysis. However, periodic examinations of the same coupons may not be reliable because deposits are removed during each inspection. Multiple coupons will avoid this problem.

Electrochemical corrosion monitoring techniques can be useful in identifying the onset of corrosion. These must be combined with bacte-

Table 10.2
Corrosion by Microorganisms (After King et al.[6])

Type of Microorganism	Growth Requirements	Corrosion and Related Problems
Algae	Light, water, nitrogen and phosphorous, CO_2, trace elements	Growth in ponds, cooling towers. Provide organic material for more corrosive organisms. Fouling of heat exchangers; differential aeration.
General aerobic microorganisms e.g. *Aerobacter*, yeasts, molds	Water, carbon source, nitrogen and phosphorous, trace elements	Differential aeration cells. Heat exchanger fouling.
General anaerobic microorganisms	Water, carbon source, nitrogen and phosphorous, trace elements	Produce organic acids which can preferentially chelate specific alloying elements.
Iron-oxidizing bacteria	Water, CO_2, oxygen, ferrous materials, nitrogen and phosphorous, trace elements	Oxidize ferrous (+2) to ferric (+3), cause blockage of pipes, create anaerobic conditions.
Nitrite-oxidizing bacteria	Water, nitrate, carbon source, phosphorous, trace elements, ammonia and aeration	Consume nitrite corrosion inhibitors; differential aeration cells.
Nitrate-reducing bacteria	Water, carbon source, nitrogen and phosphorous, sulfate, trace elements	Reduce nitrate, produce large quantities of organic acids.
Pseudomonas	Water, hydrocarbons, nitrogen and phosphorous, trace elements, manganese and iron, aeration	Breakdown of cutting-oil emulsions; cause emulsions of dewatering oils; acidifications of systems; skin infections. Provide material for anaerobic organisms. Differential aeration cells.
Sulfur-oxidizing bacteria	Water, sulfides or sulfur, CO_2, nitrogen and phosphorous, trace elements, oxygen	Produce up to 10% H_2SO_4. Concrete attack. Associated with waste disposal.
SRB	Water, carbon source, nitrogen and phosphorous, trace elements	Produce large quantities of sulfide. Severe localized corrosion.

riological analyses. In high sulfide environments, the resistance of the probe may fall rather than rise because of the formation of conductive sulfides on the wire or by bridging. Flush mounted resistance probes are more effective if placed at the bottom half of a piping system, which is normally continuously wetted. General fouling situations are less easily detected.

Sulfide corrosion is usually indicated with a rise in corrosion potential. This is usually associated with an increase in the pitting index. The absolute values obtained are not as important as the trends and changes that occur with time.

An electrochemical method that has proven practical is one in which the corrosion rate and the corrosion rate plus pitting index are recorded simultaneously. MIC is inferred by divergence of the two curves.

These steps are suggested to quantify the microbiological corrosion hazard to plants based on corrosion monitoring:

1. Biological sampling can be obtained using various methods. The organisms are grouped into their various types: General aerobic, acid producers, slime organisms, iron bacteria, SRB, nitrite-reducers, etc. The treatment program required depends on the type and quantity of the organisms present.

2. Check the quality of water entering the plant. The source of infection should be identified prior to the implementation of a treatment program.

3. For low levels of organisms, changes in population activity over time yield more information related to concern for corrosion than do the absolute number of organisms present. A high level of microbial growth in a cooling pond does not automatically indicate a corrosion hazard if water flow rates are high through process equipment.

4. Check dead legs, bypass loops, or other stagnant zones for SRB and other bacteria.

5. Good housekeeping is probably the most important means to reduce microbial problems. Sludge, debris, etc. should be removed from ponds or storage vessels; chemical treatment alone is usually not successful in a dirty system.

6. Electrochemical and other corrosion monitoring methods may be useful when combined with microbiological analyses. However, reliable prediction can only be made after a careful historical study. A test program based on intermittent corrosion monitoring only is unlikely to give the information necessary for decisions.

7. Scrutinize corrosion inhibitors carefully; some may reduce conventional corrosion but exacerbate microbial problems. The converse is also true.

MIC Control

The appropriate MIC control methodology depends on the materials of construction, the materials' existing condition, and water quality requirements. The methods are chemical treatment to obtain disinfection, physical methods to clean and remove slime and nodules, and materials selection to obtain more corrosion resistance.

Disinfection techniques that can be applied to new and existing systems are:

New: Biocide (soak/slug).

Existing: Hydroblasting followed by biocide treatment, acid cleaning, and biocide (soak/slug).

Disinfection of systems/components is required because many system materials are susceptible to MIC if bacteria are allowed to enter, colonize, and produce nodules. The presence of nodules indicates a potential for corrosion. In the 300 series SS and carbon steel materials, corrosion occurs beneath nodules. No disinfection or combination of disinfection techniques is 100 percent effective in killing all bacteria. In practice, reducing the bacteria levels to low values (< 1,000 counts/mL) has been successful in delaying colonization and subsequent MIC.

Physical Methods

Physical methods include brushing, pigging, hydroblasting, use of sponge balls or brushes, or other such devices. These methods will remove scale, deposits, and slime. They also may tend to expose or open existing corrosion sites, i.e., pits, crevices. If such physical treatments are followed by rinsing with water that contains an effective antimicrobial agent, there is a high probability that the existing MIC can be completely stopped. It is important that the method employed removes most, if not all, slime, scale and other materials—including in the pits—because the material and bacteria remaining in a pit can perpetuate corrosion.

If used repeatedly, physical methods are effective in keeping bacterial slime and deposits from forming on the surface of materials. These methods should be used in conjunction with a water treatment program designed to minimize the input and activity of organisms into the system. Inaccessible areas such as socket welds, stagnant regions, and crevices are difficult to clean by physical methods. These inaccessible areas can harbor organisms and represent the source for continued bacterial infection of the system.

Chemical Treatment

In order for biocides to be fully effective, the system should first be cleaned. Cleaning is important because corrosion products, biofilms, and debris can prevent biocides from contacting bacteria. Typical chemicals are acids and EDTA. This treatment involves downtime and requires flushing. The choice of control methods for chemical cleaning must be carefully determined to avoid damaging the system. Biocides should be used when refilling the system. Corrosion inhibitors may also be required.

Biocide Treatment

Initial water treatment using biocides is an effective method of preventing the establishment of bacteria colonies in raw cooling water, circulating water, and fire protection systems. Water treatment is not as effective in areas where bacteria colonies have become established because the biocide has difficulty penetrating the nodules. In SS systems, virtually any biocide may be used, and general corrosion due to the biocide is minimal compared to the pitting corrosion due to bacteria. However, overchlorination can also cause corrosion of SS.

Demineralized and nuclear reactor water systems are typically fabricated from SS and are susceptible to MIC. Demineralized water is relatively pure but not sterile. Water treatment with biocides is not desirable for demineralized water systems because of the changes in water quality which can result. These systems can be sterilized effectively using hydrogen peroxide at dosages as high as 200 ppm for 24 hours, ozone, or ultraviolet light. Hydrogen peroxide decomposes to oxygen and water, ozone can transform to oxygen, and none of these significantly affect water quality.

In stagnant carbon steel systems, a nonoxidizing biocide with pH adjusted to 10.5 can be an effective method to control bacteria. The nonoxidizing biocides have longer decay periods than the oxidizing biocides, and the high pH inhibits reproduction of the organisms. As an additional benefit, the nonoxidizing biocides generally produce lower corrosion rates in carbon steel than the oxidizing biocides.

Using pH control in conjunction with a biocide is an alternate method of MIC inhibition in closed loop and lay-up systems. The pH of the process water should be maintained at or above 10.5. As discussed above, monitoring of both pH and biocide residual, as well as biofilm development, are equally applicable to closed systems.

In all cases, qualification tests should be performed to assure that the biocides used will not increase corrosion of system materials. Corrosion tests are required to assure that implementation of a biocide will not aggravate pitting, crevice, and galvanic corrosion susceptibility of system materials. High dosages of oxidizing biocides can produce increased corrosion rates of some system materials at the dosage and frequency required to achieve the desired level of disinfection. These tests should be performed at an operating plant in test loops prototypical of system operation. Stagnant, low-flow, and full-flow environments should be evaluated, and the materials should be as prototypical of system materials as possible. Test coupon and linear polarization measurements can be used to evaluate corrosion.

Typical water treatments are presented in Table 10.3. With all water treatments disinfection down to 10^3 bacteria counts/mL can be obtained.

Chlorination is an effective chemical treatment, but high dosages (2 to 10 ppm) are usually required. Another method uses a targeted chlorination design that produces a local high-dosage, short-duration injection of chlorine, or any other biocide at a targeted section of a component.[2] With this technique, high chlorine levels are achieved locally, while the chlorine residual in the discharge effluent will be below the environmental limits.

An intermittent ozonation system consists of an ozone generator, an absorption tower packed with silica gel, and an injection system. A typical ozone dose is about 0.2 ppm. This concept is innovative in that the ozone generated is absorbed in the silica gel at -35°C (-31°F). To release the ozone, the gel is warmed, and the ozone is then injected into the water. This technique reduces the cost of ozonation by an order of magnitude, making ozonation a viable technique for MIC control in water systems. Ozone can degrade nonmetallics. This should be evaluated prior to implementation.[1]

Hydrogen peroxide has been effective in killing anaerobic bacteria. This treatment can require

TABLE 10.3
Typical Water Treatments

Water Treatment	Operational Limits	Limitations/ Precautions	Site/System Technical Data Required	Comments
		Oxidizing Biocides		
Chlorine	pH 6.5 to 7.5, EPA discharge limits	Continuous treatment in addition to slug feed required for bacterial inhibition	Seasonal variations in bacteria content, pH, water/system temperature, suspended solids, organic matter, chlorine demand, ammonia, iron and manganese levels	Can be used with multi-injection points to maintain TRC
Chlorine Dioxide	pH 6 to 10.0, EPA discharge limits	Gas is extremely volatile and is easily stripped from water systems	Same as chlorine	2½ times more effective than chlorine for slime control
Hypobromous Acid	pH up to 8.7, EPA discharge limits	Continuous treatment in addition to slug feed required for bacterial inhibition	Seasonal variations in bacteria content, pH, water/system temperature, suspended solids, organic matter, chlorine demand, ammonia, iron and manganese levels	
Hydrogen Peroxide	None	Requires 200 ppm dose and long contact time, ~24 hours	pH, water/system temperature, suspended solids, organic matter, chlorine demand, ammonia, iron and manganese levels	Non-polluting; applicable to stagnant systems
Ozone	Oxidizes iron and manganese, causing deposition of inorganic material	Plastics/rubber gaskets can become brittle after prolonged exposure to ozone	Seasonal variations in bacteria content, pH, water/system temperature, suspended solids, organic matter, chlorine demand, ammonia, iron and manganese levels	Applicable to stagnant systems
		Nonoxidizing Biocide		
Cationic Surfactants	Inactivated by high suspended solids; requires large dosage to be effective	Can produce minor crevice corrosion in stainless steel	Seasonal variations in bacteria content, pH, water/system temperature, suspended solids, organic matter, chlorine demand, ammonia, iron and manganese levels	Useful in closed and low-flow systems
		Nonchemical		
Ultraviolet Radiation	Low flow	Requires microscreening (20 microns)	Suspended solids	Can be used in demineralized water & low-flow systems; no byproducts

NOTE: Biocides with dispersants may be used in SS systems. The dispersant can increase the corrosion rate of carbon steel.

dosages as high as 200 ppm for periods as long as 48 hours.[1,4,8]

Use of biocides in water systems can allow the more resistant organisms to remain by eliminating their competitors. Water service companies overcome this by periodically changing the biocide. There are many commercial biocides available, and each should be reviewed carefully for its biocidal effectiveness, cost, and effects on other materials and components in the system.

pH Control

In closed or stagnant systems, pH control can be used to reduce the reproductive capability of bacteria, once a low level (less than 1,000 counts/mL) of bacteria is achieved in the bulk fluid. A minimum pH of 10.5 is required.

Corrosion Inhibitors

Inhibitors such as nitrites and phosphorous can become food sources for bacteria. These inhibitors should not be used in closed loop recirculating systems.

Remedial Measures

Design changes, such as increasing fluid velocities to reduce the bacteria residence time in the system and the elimination or reduction of crevices, stagnant and low flow areas, are effective. For closed systems infected with bacteria, the normal recommendations are to clean the system to remove deposited slime and debris and flush the system with water containing biocides. The system can be treated with a "shock" dose of biocide. For small systems, a continued low running level of biocide is recommended. Cost can become an important factor for large systems; in these, periodic shock doses are usually recommended.

System design factors such as the presence of crevices and flow rate are very important. The bacteria can enter these crevices, and corrosion may result. The use of demineralized or deionized water during testing, cleaning, or flushing does not prevent bacteria from entering the system. When the system is stagnant or when evaporation occurs, nutrients and organisms can deposit on metal surfaces and encourage bacteria to settle in those locations. These deposits can shield the bacteria from biocides or pH control additives; therefore, they should be removed.

In closed and open systems, stagnant areas should be minimized to the extent practical. This can be accomplished by installing a recirculating bypass loop to maintain water velocity.

Linings

Nonmetallic lining materials can be used to isolate the surface from bacteria, thereby eliminating MIC attack. The lining material should be chosen carefully because some pigments and resins may become nutrient sources for bacteria. Detailed application and inspection criteria are required to obtain a lining free of pinholes. Nontoxic linings may be successful.[4] These nontoxic linings have low surface energy and coefficients of friction to impede the adherence of slime and organisms. These surfaces can also be easily cleaned.

Linings by themselves cannot be expected to give complete protection against corrosion because they are permeable to water, oxygen, ions, and they might degrade or blister with service. In some cases, biocides can be introduced in linings (Table 10.4). The effects on corrosion are presented in Table 10.5.

In many cases, the linings themselves can be attacked by bacteria.[1] Microbes can attack the pigments and other additives in linings by using them as a food source. This process can produce partial degradation of the lining, resulting in breaks.

Lining materials that appear to perform well are the coal tar epoxy-polyamide enamels. Using inorganic silicates as a filler will further improve their performance.

Microbiological damage to plastics is well-known. The more polymerized the material, the more resistant it is to bacterial attack. Polystyrene, polyvinyl chloride (PVC), and polytetrafluoroethylene (PTFE) are relatively stable. However, their additives and coloring agents may not be. Softeners added to polymers can act as food sources for bacteria. This will result in severe degradation of softened plastics such as PVC tapes. PVC without additives is not normally used as a lining.

Wet Lay-up

Use of shock dosages of an approved biocide during lay-up can prevent bacteria reproduction and colonization. For systems such as the fire protection system, whose normal operating mode is a lay-up condition, pH greater than 10.5 can reduce the likelihood of MIC. Biocide additions, together with pH control, can be more effective. Care should be taken to ensure that treated waters enter the system during normal system flushing and testing. Bacterial monitoring is required to assure that the bacteria population is controlled.

TABLE 10.4

Biocides Used in Coal Tar Pitch/Epoxy Lining (After Williams et al.[3])

Biocide	Biocide Type	*Weight%
Arquad B. 100 Armohib 25 Arquad T2C-50 Crodazoline 0 CTAB Duomeen 0 Duomeen C Duomeen TDO Morpan CW Morpan TPB	Quaternary ammonium compound	1
Phelam	Organic mercury compound	1
Pentachlorophenol	Phenol derivative	5
Hibitane	Phenol derivative	0.5
Panacide	Phenol derivative	1
Zinc chromate	Chromate	20
Busan 11M1	Borate	20
Anticide APA	Fluorinated sulfonamide	2
TBTO	Organic tin compound	5
Triphenyl lead acetate	Organic lead compound	1

*Typical concentration in lining

TABLE 10.5

Reduction in Corrosion of Mild Steel by Introducing Biocides into a Pitch-Epoxy Lining

Additive	Corrosion (mg/dm^{-2}) In Culture of SRB	Corrosion (mg/dm^{-2}) In Sterile Medium	Corrosion Relative to Standard Coating in Culture of SRB, %
Panacide	26.5	2.7	30
Zinc chromate	29.9	3.4	33
Hibitane	35.0	3.2	39
Crodazoline O	43.3	4.0	48
Morpan CW	44.2	4.0	51
Morpan TPB	44.9	4.1	51
TBTO + Arquad B.100	46.3	4.7	54
Duomeen O	48.2	2.7	54
Armohib 25	49.5	3.8	57
Arquad 2C 50	49.6	3.4	57
Triphenyl lead acetate	52.0	3.5	60
Busan 11-M1*	56.0	1.4	65
Zinc chromate	59.1	2.8	66
CTAB	61.0	3.2	69
Arquad 2C 75	61.5	2.6	69
Phelam	62.7	3.7	72
TBTO	63.6	3.5	72
Arquad B.100	63.8	0.1	72
Duomeen TDO	64.1	4.4	72
Duomeen C	68.8	3.9	78
Busan 11-M1	73.9	3.6	87
Acticide APA	81.0	3.5	93
Pentachlorophenol	85.1	2.1	96
None	87.9	4.0	100

Cathodic Protection

Impressed-current cathodic protection, used in conjunction with lining materials, will provide corrosion protection if the lining material becomes damaged during service. Use of linings reduces the current required. Without linings, the reliability of cathodic protection systems to protect against MIC is uncertain. Normally in these situations an additional 100 mV over that required to prevent normal corrosion may provide the necessary protection against MIC. However, the level of cathodic protection used with linings to protect components from corrosion in nonbacterial environments may not be sufficient to protect against corrosion in the presence of bacteria.

Prior to implementation, the effect of cathodic protection on other materials should be carefully reviewed. For example, titanium tubed condensers can be damaged due to hydrogen embrittlement if the potentials are too high.

Corrosion-Resistant Materials[11]

Material upgrading alone usually will not provide immunity from MIC. The use of more resistant materials should be combined with an effective water treatment program. In low-velocity and stagnant systems multiple locations for biocide injection (targeted injection) can provide the most protection against MIC. Nonoxidizing biocides, with their lower decay rates, can provide longer protection in stagnant or low-flow areas.

Onsite corrosion tests should be representative of the materials used and the environments tested. The effect for low-flow, stagnant, and normal-flow environments should be evaluated. Crevice, pitting, and galvanic corrosion tests should be performed. These tests should be long enough to assess the effects of seasonal variations in bacterial content.

The materials processing and test specimen geometry should be prototypical. Typical metallurgical concerns are material cleanliness (inclusions), chemistry (particularly sulfur), heat treatment (annealing, stress relieving, welding effects), machining (such as weld counterbores), bending and pickling, and passivation requirements. Welding variables to be evaluated include heat input, preheat, joint design, and the metallurgy of the heat-affected zone.

Prototypical pipe spools with welds can be tested in service using water sidestream loops, which receive the same chemical treatment as the actual system. Field tests should be performed for one year to expose the materials to environmental and biological seasonal changes, and to ensure that tests are performed under worst-case conditions. The contribution of MIC can be evaluated using a biocide source that will not change the water chemistry, such as UV light.

The use of more corrosion-resistant materials such as titanium, the higher alloy SS, nickel-base alloys, and alloys containing high molybdenum must be carefully examined. The effect on overall cost, an industrial infrastructure to supply all product forms, and code approvals should be evaluated. Though these materials may be more resistant, they may not be immune. For example, MIC of SS welds may be reduced by use of nickel-base filler metals or solution annealing. However, use of the nickel-base alloy filler metals may shift attack to the weld heat-affected zone or base metal. Solution annealing of field welds is obviously impractical. Material upgrades can provide more margin against MIC if bacterial colonies develop.[7]

Repair measures:

- Use of full penetration welds to eliminate the crevices produced by socket weld joints.
- In carbon steel systems, leaking areas can be repaired by weld overlay, use of clamps, structural sleeves, and rubber boots. If complete spool replacement is required, the use of more corrosion-resistant materials should be considered in conjunction with a reliable, consistent water treatment plan and an evaluation for the potential for galvanic corrosion.
- Weld overlay in carbon steel systems.

Without adequate water treatment to control corrosion influencing bacteria, all these repair techniques are only temporary. The use of higher alloy weld filler metals or full penetration welds require grinding out the existing weld deposits.

The use of structural sleeves, clamps, and rubber boots have the least impact on plant operations, but may also be the least permanent of possible solutions.

Regardless of the repair technique used, surveillance inspections should be performed. In SS systems, visual and/or radiographic methods can be used and should include both the weld deposits and heat-affected zones. In carbon steel systems, appropriate NDE methods are ultrasonics and radiography.

Special Areas

Corrosion of Steel Pilings in Port and Harbor Loading Areas

Over the years, studies have indicated that the corrosion of sheet steel piling can be conveniently divided into five zones:

- Atmospheric,
- Splash,
- Intertidal,
- Immersed, and
- Sediment.

Studies in the United Kingdom and the West Indies indicate that enhanced corrosion, resulting from SRB activity, does occur under soft-bodied fouling after about five years in service, presumably the minimum time to establish a local ecosystem.[5] Corrosion rates were 30 to 50 percent higher than expected. The assessment procedure described in Tables 10.6 and 10.7, based on measurement of sediment corrosion, can be considered as a guideline to the exacerbation of corrosion beneath marine fouling by SRB.

Preventative Measures

For new construction, where the risk of enhanced corrosion by SRB is unacceptable, the following actions can be taken:

- Use a thicker wall material,
- Coat the piling, and
- Apply cathodic protection.

For established construction, one can:

- Use cathodic protection,
- Minimize spillage of degradable cargo or oil,
- Replenish fouled harbor waters by pumping, and
- Periodically remove decaying material (however, this may increase the aggressiveness of the sediment by producing a "compost heap" at the foot of the piling).

TABLE 10.6

Assessment of Enhanced Corrosion by SRB in Harbor Sediments (After Ronay et al.[9])

Factor	**Assigned Points**
Nature of Sediment	
Mud	2
Sand, minor mud	0
Organic content (Wt. % basis)	
If mud > 15%	4
10 to 15 %	3
5 to 10 %	2
< 5 %	0
If sand	0
Availability of nitrogen and phosphorous	
If high	
(a) high organic content	3
(b) medium organic content	2
(c) low organic content	0
If low	0
Temperature °C	
> 15	3
5 to 15	2
< 5	0
Sulfate content (ppm)	
> 1,000	3
500 to 1,000	2
200 to 500	1
< 200	-1

NOTE: For a particular harbor sediment, the points are totalled, and from the numerical value, the expected corrosion rate is presented in Table 10.7. The assessment procedure can also be used to estimate the risk of corrosion under fouling, but the points do not relate to corrosion rates.

TABLE 10.7

Assessment Points and Sediment Corrosiveness (After Ronay et al.[9])

Points	Corrosion Rating	Estimated Sediment Corrosiveness
10 to 15	Enhanced attack	100 to 500 µm/yr
5 to 10	Localized damage	30 to 100 µm/yr
0 to 5	"Normal" rates	0 to 30 µm/yr

Microbiological Degradation of Lubricating Oil[10]

All infections occur after water enters the oil system. The volume of water needed for bacterial growth is extremely small in proportion to the total volume of lubricating oil. If lubricating oil could be maintained dry, the problem of microbial infection would not arise. However, even assuming that no leakage occurs, the water content in the oil may rise with time due to condensation; this can also occur in oil storage tanks. In some cases, the design of lubricating systems can prevent water from being easily removed, if at all, without a major and costly effort. Water treatment with nitrite and phosphate inhibitors to reduce corrosion of metal surfaces supplies the microbes with essential growth materials. Nitrites are themselves usually biocidal, but several organisms can reduce them to ammonia. In addition, they degrade to ammonia by reacting with the steel surface when inhibiting. Alternatively, the nitrite oxidizes to nitrate, which is nontoxic to microorganisms.

Remedial Action

In extreme cases, oil replacement is probably the only solution. Drained oil should be discarded remote from other metal and aqueous areas, and it should not be mixed with or contact new oil. If allowed, burning of oil is a suitable disposal method.

The system should be thoroughly cleaned by steam lancing of walls and structures to remove slime and deposits followed by opening of the sumps and scouring out deposits. The most practical way to clean oil lines is to flush them out using a light flushing oil. The flushing oil can be dosed with biocide. Oil storage tanks should also be inspected frequently and cleaned if necessary.

Water Side

Because the primary source of infection may be located in the cooling water jacket, the water system should be drained and filled with freshwater containing a biocide. The treated water should be allowed to stand in the system for a minimum of 24 to 48 hours. Attention should be paid to blanked lines and bypass loops to ensure that they are adequately cleansed. The system should then be drained, and the water and slime discarded.

The water used to replenish the system after draining should contain a biocide to prevent infection. If an inhibitor is used, its selection should be carefully considered. Inhibitors low in phosphate and nitrate are recommended because they can become food sources for *Pseudomonas*. Some corrosion inhibitors contain biocides, or biostats (they prevent growth rather than kill microbes). It is important that the user knows the type of bacterial control being used.

Oil Additives

As a class of compounds, benzyl cresols have been used at concentrations of up to two percent. Other aromatic biocides can be used, but attention to the particular oil characteristics is necessary because not all additives will be compatible with all oil formulations. The biocide must pass into the water phase to be effective. In engines that run "wet," the biocide will be exhausted fairly rapidly.

The use of biocides in oils is probably most effective for long-term storage rather than in the engine system. Sumps with low in-leakage of water would benefit from this type of treatment.

References

1. A.A. Stein, "Microbiologically Induced Corrosion," presented at the Electric Power Research Institute (EPRI) Workshop on Microbe-Induced Corrosion in Nuclear Power Plants in Palo Alto, CA, 1986.
2. "Microbiological Corrosion of Sheet Piping," The Corrosion and Protection Centre Industrial Services (Manchester, UK: University of Manchester, Institute of Science and Technology [UMIST]).
3. M.E. Williams, R.A. King, J.D.A. Miller, "Sulfate-Reducing Bacteria Surface Coatings and Corrosion," *Journal of the Oil and Colour Chemists' Association* (1973).
4. A.A. Stein, Y.G. Mussalli, "Microbiologically Induced Corrosion Control in Cooling Water Systems," presented at the 1986 Joint Power Generation Conference in Portland, OR.
5. R.A. King, J.D.A. Miller, "Corrosion of Ferrous Metals by Bacterially Produced Iron Sulphides and Its Control by Cathodic Protection," presented at the First International Conference on the Internal and External Protection of Pipes at the University of Durham, 1975.
6. R.A. King, "Monitoring of Microbiological Activity and Its Relation to Corrosion," Corrosion & Protection Centre (Manchester, UK: UMIST).
7. A.A. Stein, "Microbiologically Induced Corrosion," presented at EPRI Workshop on Microbe-Induced Corrosion in Nuclear Power Plants in Palo Alto, CA, 1983.
8. A.A. Stein, "Identification and Counteraction of Microbe-Induced Corrosion in Metallic Systems," *Chemical Processing* (1986).
9. D. Ronay, I. Fesus, A. Wolkober, "New Aspects in Research in Biocorrosion of Underground Structures," presented at UK Corrosion '87 in Brighton, UK, 1987.
10. R.A. King, P. McKenzie, "Microbial Degradation of Marine Lubricating Oil," *The Institute of Marine Engineers* (London, UK: 1976).
11. A.A. Stein, "Materials Selection Criteria to Control Microbiologically Influenced Corrosion in Once Through Systems," *Microbial Corrosion 1988 Workshop Proceedings*, EPRI ER-6345 (Palo Alto, CA: EPRI, 1988).

Chemical Treatment for Mitigation of MIC

By A.A. Stein and Y. Mussalli

In evaluating a chemical treatment program, the following issues should be evaluated for each treatment:

- Disinfection characteristics related specifically to microbiologically influenced corrosion (MIC),
- Regulatory constraints,
- Corrosion potential, and
- Costs.

Typical treatments

Hypochlorous Acid

Hypochlorous acid is the most commonly used biocide in cooling water systems and one of the most powerful oxidizing agents. The sources of hypochlorous acid are chlorine gas and sodium hypochlorite. At power plants, sodium hypochlorite is used because it is considered safer than gas. Sodium hypochlorite hydrolyzes to hypochlorous acid (HOCl) and sodium hydroxide. Hypochlorous acid also dissociates into hypochlorite (OCl^-) ion. The amount of hypochlorite ion determines the biocidal killing ability. The pH of treated water determines the extent of ionization of hypochlorous acid.

The formation of hypochlorous acid is most favorable in waters with a pH value between 6.5 and 7.5. At pH values greater than 9, hypochlorous acid becomes ineffective.[1] Sodium hypochlorite injections to produce 0.2 to 0.5 ppm residual chlorine are not considered harmful to most materials.

Hypochlorous acid reacts rapidly with inorganic reducing agents such as sulfides, sulfites, and nitrites, and with some dissolved organic materials. The reaction with ammonia and some organic compounds produces halogenated compounds (chloramine) and renders hypochlorous acid less effective.

Hypochlorous acid can react with iron and manganese. The deposits produced can result in under deposit corrosion, especially of stainless steels (SS).

Biocidal action is the result of the rapid oxidizing power of hypochlorous acid. Bacterial cells are destroyed by the irreversible chemical reaction of hypochlorous acid with a host of cellular components.[1] Hypochlorous acid is effective in controlling slime.

Sodium hypochlorite is currently regulated by the U.S. Environmental Protection Agency (EPA) to 0.2 ppm total residual chlorine (TRC) in the effluent for two hours per day. At this level, it can control planktonic bacteria, but it is less effective in controlling sessile microbes that contribute to MIC. Future changes in the Water Quality Act in water quality distribution limits may lower chlorine to 10 ppb. These levels are almost undetectable and will not be effective in controlling planktonic bacteria.

Targeted chlorination and multiple-injection techniques were developed recently for better use of chlorine as a disinfectant.

Targeted Chlorination

Targeted chlorination is a sequential pulse injection of high dosages of chlorine (sodium hypochlorite) to one section at a time. It was applied to a condenser tubesheet until all the tubes were treated.[2] Dosages were 16 to 20 ppm

Table 11.1

Effects of Chlorination on Materials in Seawater

Material	Chlorination Level	Effect
Al-6X	2 ppm continuous	Crevice corrosion after 3 months exposure
	10 ppm 4 hr/day	Crevice corrosion after 3 months exposure
	0.5 to 0.8 ppm	No crevice corrosion
316 SS	0.1 to 0.2 ppm	No corrosion
Copper	3 ppm intermittent	Corrosion
Carbon Steel	1 ppm continuous	No corrosion

NOTE: Corrosion of copper alloys may result with intermittent free chlorine levels of 3 ppm at high velocities. At lower velocities, chlorination may lower the corrosion rates.

for 3 min, once a day. Accelerated corrosion evaluations were performed on several materials including titanium, AL-6X, Cu-Ni alloys, admiralty, and SS.[3,4] Pulse dosages of up to 20 ppm for up to 15 min, three times a day, did not result in adverse effects on corrosion of the tube and tubesheet materials.[3,5]

Electric Power Research Institute (EPRI) conducted an extensive literature search on the effects of chlorination on tube and tubesheet materials.[5] Table 11.1 presents information on increased corrosion of various materials by chlorination.

In summary, most common alloys can be adversely affected by chlorination, depending on dosage level and frequency. Depending on the circumstances, chlorination could be beneficial to corrosion rather than cause a detrimental effect.

Targeted chlorination can produce several benefits:

- Reduction of sodium hypochlorite consumption by up to 80 percent,
- Improvement in cleanliness and heat transfer capability of condensers, and heat exchangers by up to 20 percent (Figure 11.1), and
- Reduction of total residual chlorine (TRC) levels in effluent to below 0.1 ppm (Figure 11.2), and reduction in halogenated compounds (trihalomethanes [THM]) to very low levels (Figure 11.3).

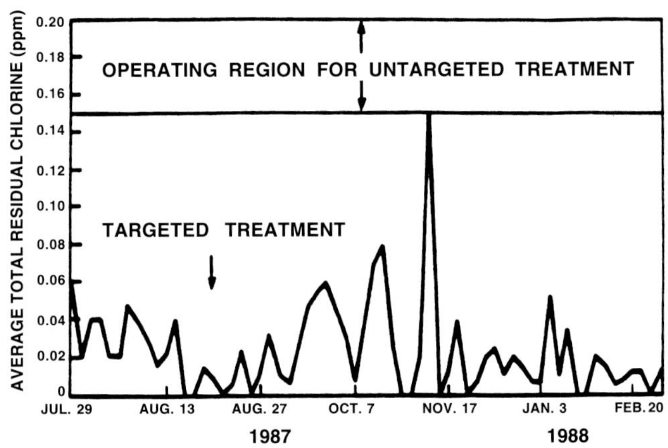

FIGURE 11.2
Total residual chlorine at the effluent of the targeted and conventional treatment.

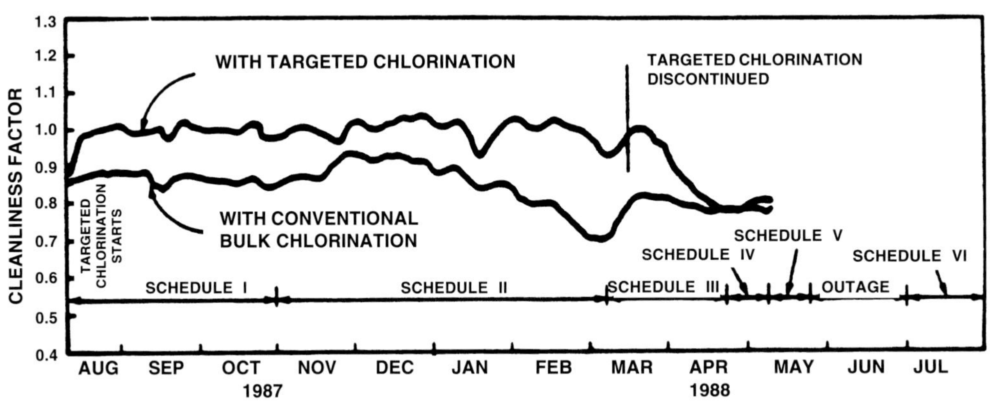

FIGURE 11.1
Cleanliness factors for targeted vs untargeted treatment.

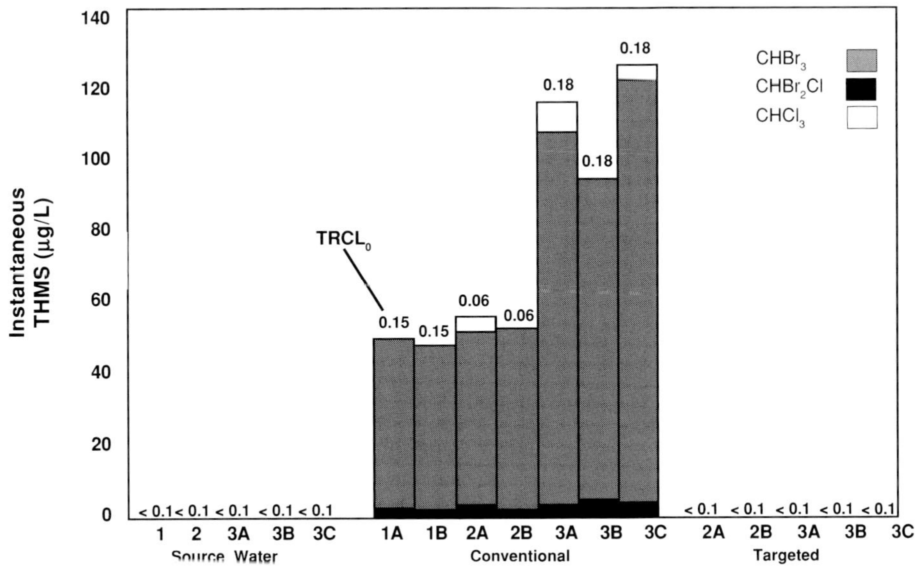

FIGURE 11.3
Instantaneous trihalomethane formation for conventional and targeted chlorination.

Targeted techniques are beneficial to service water systems and MIC control because:

• The technique can be easily adapted to the service water piping system through multiple injection of biocide into parallel pipe loops and upstream of a series of heat exchangers. In this manner, a higher level of biocide can be applied to the pipes or heat exchangers. Also, dilution by the untreated loops can result in low levels of residual biocide at the effluent of the service water system.

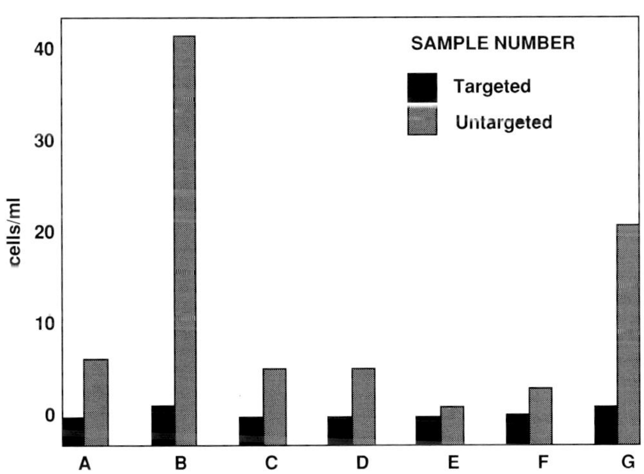

FIGURE 11.4
SRB counts.

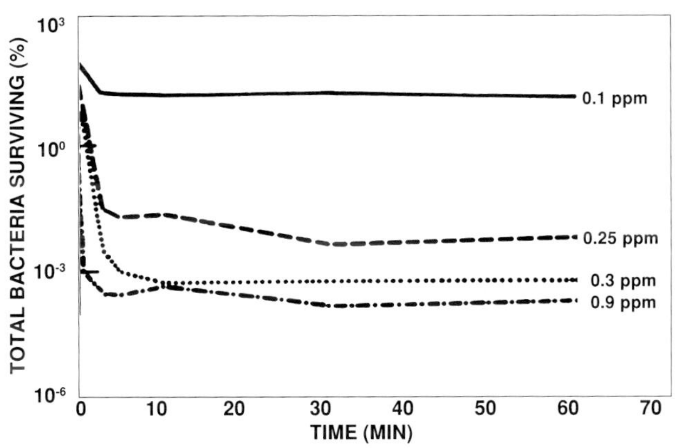

FIGURE 11.5
Seawater disinfection curves.

• Targeted chlorination, with its higher free available chlorine (FAC) dose, can remove slime and sulfate-reducing bacteria (SRB) that usually grow under the slime layer. Figure 11.4 presents the beneficial effects of targeted chlorination over bulk chlorination on SRB in a seawater-cooled condenser. Its effectiveness is due to the much higher disinfection capabilities of high FAC dosages. For example, a dose of 1 ppm can kill 1,000 times more bacteria than a dose of 0.1 ppm (Figure 11.5), even though the dose level only increases by one order of magnitude.

• The intermittent pulsed injection may not result in adverse corrosion of the common materials used in service water systems.

Other works that support these findings are:

• Miller & Bott found that a concentration of 4 ppm FAC resulted in four to six times higher film removal than at 2.5 ppm, and a higher velocity of sodium hypochlorite solution over the biofilm resulted in a higher rate of removal.[5]

• Scott and Davies evaluated MIC on high molybdenum stainless steel and recommended a shock treatment of 1.2 to 1.5 ppm FAC for one hour per day at the heat exchangers to remove sessile bacteria.[4] Maintenance levels of 0.10 ppm were used to control other fouling organisms. The reason for the low levels was to permit the repassivation of the film on the metal surfaces and to prevent the development of chlorine-tolerant strains.

• Characklis monitored the decay of chlorine in the service water system of a nuclear power plant.[6] He found that free available chlorine decays within 50 ft of its point of injection in the service water pumps. The piping was 800 ft long, and injected dosage was 3.54 ppm. Therefore, multiple injections along the system and upstream of heat exchangers can be advantageous.

Chlorine Dioxide

Chlorine dioxide ($Cl0_2$) has more oxidizing power than chlorine and is less affected by the chemical composition of the water than is chlorine. Complex organic molecules and ammonia do not react with $Cl0_2$, which is also pH insensitive. $Cl0_2$ is used in cooling waters contaminated with ammonia and phenols. $Cl0_2$ does not form halogenated compounds; therefore, it is more acceptable environmentally. $Cl0_2$ does not react with water and remains a true dissolved gas. It is safe to use when handled in a water solution. However, the gas is extremely volatile; therefore, it is easily stripped from water systems with small amounts of aeration.

$Cl0_2$ is usually generated on site by the reaction between sodium chlorite and sodium hypochlorite in the presence of sulfuric acid to ensure proper pH. $Cl0_2$ is generally more costly than chlorine.

The following case studies document the beneficial uses of $Cl0_2$ in controlling and reducing total bacterial counts in water:

• Sussman and Ward reported on control of SRB at 1,200 colonies/mL and plate counts of 1,000,000/mL.[7] The treatment with $Cl0_2$ for four hours per day supplemented the shock dose of nonoxidizing biocide to reduce bacterial count. In other cases, $Cl0_2$ was more effective in controlling SRB than chlorine.

• Ren-Hao et al. reported that $Cl0_2$ as a biocide is superior to chlorine in controlling SRB, slime, and iron bacteria.[8] Dosages of 0.5 ppm for 30 min killed 98 percent of SRB. $Cl0_2$ was 2.5 times more effective than chlorine in controlling slime. Data showed that 90 to 95 percent kill rate can be achieved at 2 ppm $Cl0_2$ against slime-forming bacteria and fungi, and 1 ppm against iron bacteria and SRB. They are considering using $Cl0_2$ in thermal power plants.

• Ward et al. presented a case study of a chemical plant that used $Cl0_2$ at 1.2 ppm for three hours per day instead of chlorine.[9] Substantial reductions in nonoxidizing biocide addition were achieved. The system indicated corrosion rates of only 0.5 mils per year (mpy) on mild steel.

Hypobromous Acid

Recently, hypobromous acid has been replacing hypochlorous acid in water with pH >8 as a disinfectant. Claimed advantages include:

- Active over a broad spectrum of bacteria,
- Lower corrosion rates than chlorine, associated with reduced required dosages,
- Achieves fast kill rates,
- Decay rates of free and combined bromine compounds have half lives of 10 to 60 min, and
- Complies with current EPA limits of 0.2 ppm total residual oxidant.

Bromine is available in several forms:

- Bromo-chloro-dimethyl-hydantoin (BCDMH),
- Activated Bromide (Br^-), and
- Bromine Chloride (BrCl).

In all cases, the principal mechanism underlying its biocidal activity is the formation of hypobromous acid, HOBr.

Several papers discuss bromine chemistry and case histories:

• Conley et al. presented disinfection curves comparing hypochlorous and hypobromous acids in controlling several types of bacteria.[10] Hypobromous acid achieved kill levels of four orders of magnitude higher in less than four min.

• Fellers et al. presented data on the reduced Cu-Ni corrosion by bromine and on its potential to reduce manganese deposition and copper-alloy corrosion.[11] Bromine was capable of reducing manganese deposition for under-deposit corrosion and MIC control.

• Bongers et al. evaluated hypobromous acid at an estuarine power plant with low salinity and found that a continuous dose of 0.5 ppm can control biofouling.[12]

• Burton et al. concluded that bromine residuals may be less toxic to aquatic organisms than chlorine residuals.[13]

Hydrogen Peroxide

Hydrogen peroxide (H_2O_2) is a relatively strong biocide and disinfectant, less effective than ozone but equivalent to hypochlorous acid. It is easy to apply and is nonpolluting. Very little large-scale use of hydrogen peroxide has occurred, especially with respect to disinfection applications.[14] Hydrogen peroxide requires large doses with long contact time to be effective. It is chemically an unstable liquid, safe to handle, but can be corrosive to many conventional engineering materials.

Wilde et al. found that concentrations of 1 ppm can increase the susceptibility of aluminum to corrosion.[15] H_2O_2 is used as a corrosion inhibitor in systems containing SS.

Ozone

Ozone is the strongest oxidant that can be safely used in water treatment. Ozone is an extremely unstable gas with a half-life of 20 to 30 minutes, thus requiring rapid mixing in the water stream. Due to the reactivity of ozone, it must be generated on site. Ozone is a highly toxic and dangerous gas, and it requires large storage facilities due to the lack of compressibility. However, ozone is considered safer than chlorine.[14] Work done on ozonation of tertiary effluent revealed some corrosion of type 304 SS (UNS S30400).[16] EPRI evaluated the use of ozone as an alternate to chlorine.

The results at a freshwater site indicated that dosages of 1.0 ppm of ozone are required, compared to 0.1 ppm TRC of chlorine. The ozone oxidized manganese to an insoluble form leading to deposition of inorganic materials via coprecipitation.

EPRI found that ozone, and to a lesser degree chlorine, chlorine dioxide, and bromicides were effective in controlling organisms involved in MIC.[17] The lethal effect on bacteria is extremely rapid, requiring 5 to 10 seconds of contact time for total control. A residual of 0.02 ppm ozone achieved total disinfection.

IBM conducted an ozonation pilot study for a cooling tower.[18] Treatment residuals were 0.1 ppm. Coupons of mild steel, copper, and aluminum were monitored for corrosion. Corrosion rates in mpy were 1.0, 0.02, and 0.02 for steel, copper, and aluminum, respectively. Ozone can come out of solution in heat exchangers, forming bubbles that reduce heat transfer.

Only a few materials are resistant to prolonged contact with ozone. They are: SS, aluminum, fluorinated hydrocarbons, ceramic, and glass. For flexible piping, only fluorinated plastics (e.g., polytetrafluorethylene [PTFE]) are suitable. For gasket materials, ethylene-propylene elastomers can be used. Other plastics or natural rubbers are only partially resistant. These materials become brittle after extended contact with ozone and can result in water leakage with concurrent release of ozone.

Electrolytic Techniques

The electrolytic production of sodium hypochlorite is widely used. The Central Electricity Generating Board of the United Kingdom has standardized the use of electrochlorination. In the US, several seawater and freshwater power plants use electrochlorination. The electrode pack arrangement consists of:

- Stable anodes and titanium or nickel alloy cathodes in a monopolar configuration and
- Low-resistance, pure copper conducting bars protected by corrosion-resistant titanium tubes.

This system has a small constant anode-cathode gap for low power consumption, and it can be combined into electrically parallel stacks of one to four, thus allowing for various system design capacities. The process electrolyzes brine to produce sodium hypochlorite in a safe and efficient manner. In the process, hydrogen gas is produced, which can be diluted with air or water for safe discharge.

The electrolytic techniques have potential for use in the multiple-injection of biocides into the service water system. Small units using brine can be installed inside the power plant along the system to produce sodium hypochlorite or mixed oxidants for direct injection into piping. An isolation valve and a check valve will interface between piping and the biocide producing system. Therefore, the electrolytic technique may be safer than using a liquid hypochlorite or activated bromide addition to pipes inside the plant.

Ultraviolet Treatment (UV)

UV is used to disinfect drinking water. Recently, large UV units, approximately 10,000 gpm, were used to disinfect seawater for reinjection in oil wells in the North Sea.[19,22] Kills on the order of 99.9 percent of the total SRB content were achieved.

New high power, low-pressure mercury vapor lamps were developed in Europe with an output intensity many times higher than that of conventional lamps. The bactericidal radiation intensity of 165 watts allows disinfection to occur in 1 second. To accomplish the required contact time, the lamps are set in series in a pipeline or inserted in an open bay perpendicular to the flow. To be effective, the following conditions are necessary:

- The turbidity of the water should be low, although turbidity is not necessarily inversely proportional to UV transmission.
- The lamps should be clean of slime or silt to maintain UV transmission. The UV wavelength should be 254 nm.

Concerns from an operating viewpoint include:

- Life of UV lamps is limited to about eight months.
- The power requirements are 25 watts/m^3 or 90 mW/gallon.
- The biocide effect occurs only in the presence of UV light. The normal radiation dose is 25 mW/cm^2 at a wavelength of 254 nm for potable water disinfection.
- The head loss on pumps due to the lamps is approximately 5 psi.
- Although no environmental impacts are expected from UV treatment, nontarget species may also be killed.
- Water should be microscreened to 20 microns.

For these reasons, UV is more applicable to low-flow systems and demineralized water systems that can be filtered and are not open to microbial contamination downstream of the UV treatment.

Nonoxidizing Biocides

Nonoxidizing biocides are primarily used in cooling tower recirculating water systems. They are used when chlorine proves ineffective in controlling fouling because of high chlorine demand

7. S. Sussman, W. Ward, "Microbiological Control with Chlorine Dioxide Helps Save Energy," CORROSION/77, paper no. 65 (Houston, TX: NACE, 1977).
8. Ren-Hao, et al., "Study of Microbial Fouling and Its Control in Desalting Water Systems of Heat and Power Stations," *Proc. Inst. Congress Met. Corrosion* 4(1984): pp. 213-220.
9. W.J. Ward, et al., "Advantages of Chlorine Dioxide as Biocide," *Journal of Ammonia Plant Safety* 20(1978): pp. 64-69.
10. J. Conley, et al., "Bromine Chemistry, an Alternative to Dechlorination in Cooling Water and Wastewater Disinfection," *IWC-87-42.*
11. B.D. Fellers, et al., "Bromine Replaces Chlorine in Cooling-Water Treatment," *Power* (1988): pp. 15-20.
12. K. Bongers, et al., "Bromine Chloride—An Alternative Biofouling Control Agent for Cooling Water Treatment," *Water Chlorination* 2 (Ann Arbor, MI: Ann Arbor Science, 1977), pp. 735-752.
13. D. Burton, L. Linden, "Biofouling Control Alternatives to Chlorine for Power Plant Cooling Water Systems: An Overview," *Water Chlorination* 2 (Ann Arbor, MI: Ann Arbor Science, 1977), pp. 717-734.
14. Waite, et al., "Alternatives to Chlorination," *Water Chlorination* (Ann Arbor, MI: Ann Arbor Science, 1977).
15. E.W. Wilde, et al., "Selecting an Algicide for Use with Aluminum Alloys," *Water Research* 15(1981): pp. 1,117-1,124.
16. J.T. Johnson, J.T. Longaker, "A Comparison of Conventional and Unconventional Controls for Microbiologically Influenced Corrosion in Raw Water Piping," *ASME-WAM*, HTD-Vol. 86, pp. 59-62.
17. "A Study of Microbiologically Influenced Corrosion in Nuclear Power Plants and a Practical Guide for Countermeasures," *EPRI NP-4582* (Palo Alto, CA: EPRI, 1986).
18. *Ozonation as Cooling Tower Water Treatment: A Pilot Study* (Armonk, NY: BM, TR 20.0430, 1983).
19. J.B. Clark, et al., "Using Ultraviolet Radiation for Controlling Sulfate-Reducing Bacteria in Injection Water," Society of Petroleum Engineers of AIME, 1984.
20. A.A. Stein, et al., presented at Fourth International Symposium on the Environmental Degradation of Nuclear Materials—Power Reactors in Jekyll Island, GA, 1989.
21. "Dechlorination Technology Manual," *EPRI CS-3748* (Palo Alto, CA: EPRI, 1984).
22. A.A. Stein, "Service in Water Materials Test Program," paper presented at EPRI Service Water System Reliability Improvement Symposium.
23. Jose Gines, "Control of Fouling in Open Cycle Heat Exchangers," presented at the 1993 EPRI Service Water System Reliability Improvement Symposium.

MIC Case Histories

Compiled by J.G. Stoecker

This chapter presents color photographs of various forms of microbiologically influenced corrosion (MIC) in case history format. Color is an extremely important aspect in studying this type of corrosion and will help the reader identify and diagnose potential MIC problems in the field.

Each case history was selected for a specific characteristic, including MIC, that is typical of metals, alloys, and MIC problems in various industries. Some case histories are presented in their entirety when symptoms, analytical findings, probable mechanisms, and corrective measures are appropriate. Others are short excerpts from original cases and include only significant details.

The Case Histories section includes these cases:

Aluminum (AL) — One case
Carbon steel (CS) — Two cases
Mixed metals (MM) — Two cases
Nickel (NI) — Two cases
Stainless steels (SS) — Twelve cases

Effects of Fungal and Bacterial Contaminants of Kerosene Fuels on the Corrosion of Storage and Distribution Systems[(1)]

By H.A. Videla, P.S. Guiament, S. DoValle, and E.H. Reinoso

Abstract

The electrochemical behavior of pure aluminum, 2024 alloy (UNS A92024) and mild steel have been studied in the presence of several species of fungi and bacteria isolated from kerosene fuels storage and distribution systems. Microbial growth facilitates passivity breakdown through the action of organic acidic metabolites derived from hydrocarbon degradation. Their corrosive effect depends on the pH and electrolyte composition of the medium, especially with reference to chloride and nitrate levels. Microbial adhesion to the metal surface enhances metal dissolution. The metabolic activities of different microbial contaminants of fuel/water systems sometimes lead to opposite effects with respect to the corrosion of the metal. Thus, the uptake of nitrate as nitrogen source by *Hormoconis resinae* (*H. resinae*) can be an important alternative corrosion mechanism. A protective effect has been observed for aluminum and 2024 alloy in the presence of *Serratia marcescens* (*S. marcescens*) in spite of its hydrocarbon degradation capacity.

Introduction

Microbial contamination of hydrocarbon fuels is the main cause of serious problems concerning the quality of maintenance of the product, as well as the corrosion of metals used during the processes of extraction, production, distribution, and storage of the fuel.

Microorganisms usually present in droplets or thin films are sufficient to allow microbial growth and the development of microbiologically influenced corrosion (MIC). Because of the electrochemical nature of MIC an aqueous environment is required.

Fungal growth can also occur on side walls not necessarily adjacent to large waterbottoms. Hyphal growth of fungi usually ramify over the fuel/water interface becoming a site for further water entrapment.

As a consequence of this microbial propagation action, penetration and breakdown of fuel tank coatings and passive films occurs on the metal, leading to the onset of localized corrosion processes generally under the appearance of pits on the metal surface.

Due to its economic and technological importance, the microbial deterioration of jet fuel and subsequent corrosion of aluminum alloys used in the aircraft industry have been extensively studied during the past two decades.[1,2] However, little information is available on a whole interpretation of the corrosion process due to the action of microbial contaminants of the fuel. Since the late 1970s, the causes affecting the whole bioelectrochemical process have been studied in our laboratory. Thus, the following have been successively reported:

[(1)]Originally published as CORROSION/88 paper no. 91 (Houston, TX: NACE, 1988)

- Interrelationship between electrochemical and biological variables involved,[3]
- The use of localized corrosion techniques and parameters to assess and differentiate microbial activities,[4,5]
- The use of microbiological parameters like the redox-potential in conjunction with electrochemical techniques,[6]
- The analysis of the physiochemical composition of the medium in relation to corrosion,[7] and
- The importance of adhesion processes due to fungal attachment to the metal substrate.[8,9]

Some of the mechanisms proposed in the literature to interpret the MIC of aluminum in fuel/water systems are:

- Production of organic acids derived from the degradation of hydrocarbon chains,[2]
- Establishment of concentration cells at the sites of microorganisms attachment,[10]
- Formation of tubercles by the symbiotic association between certain types of bacteria and fungi,[11] and
- Direct removal of certain metallic atoms from the metal surface by extracellular enzymatic activity.[12]

Nevertheless, in the present state of understanding, the whole MIC process can only be interpreted by several mechanisms acting consecutively or simultaneously and derived from the direct action or indirect action of the microorganisms.[8] According to our publications on the subject, the MIC process would be accomplished through:

- Local increase in the proton concentration derived from organic acidic metabolites,
- An increase of the oxidizing characteristics of the medium favoring pitting attack,
- Metabolite production decreasing the surface energy of the interface passive film/electrolyte,
- Microbial adhesion processes enhancing metal dissolution, and
- Microbial uptake of corrosion inhibitors (mainly nitrates and phosphates).

The goal of this paper is to report experimental evidence on the last effect due to the uptake of a corrosion inhibitor of aluminum and its alloys by the fungus *H. resinae*. This alternative mechanism could explain some effects of fungi on the behavior of localized corrosion parameters like the pitting potential in the MIC of aluminum alloys. The corrosive action of other fungi also will be discussed.

Finally, the protective action of *S. marcescens*, one of the bacteria isolated from a jet fuel distribution system, is compared with the effect of a strain of *Pseudomonas sp.*, a usual contaminant of kerosene. Bacteria in soils and waters use paraffinic hydrocarbons in the range C10-C18 (kerosene) rather than in the range of C5-C9 (gasoline).[13] In a water-free fuel, the microbial detrimental activity is nil. As soon as water becomes available, the growth of microorganisms is possible by using the hydrocarbon as a source of carbon for their metabolism. Chemical contaminants already present in the fuel and water provide some nitrogen source and the necessary trace elements for growth. Thus, the major requirement for microbial activity is the presence of water. In the processing of hydrocarbon products, it is an almost impossible task to eliminate water. Although good housekeeping procedures could minimize the amount of water accompanying the fuel, it is possible for microorganisms to generate their own water phase for further proliferation.[14]

The use of hydrocarbon products and water in microbial activity generates small ecosystems, such as tubercles, slime deposits or biofilms on the metal surface, and all these ecosystems retain water. An assessment of the microbial ecology of fuel tanks and storage systems for aviation turbine fuel reported a wide range of microorganisms, but in almost all cases microbial contaminants were fungi and bacteria.[15,16]

The fungus *Cladorsporium resinae*, now taxonomically reclassified[17] as *Hormoconis resinae*, and some species of the genera *Aspergillium*, *Penicillium*, and *Fusarium* have been reported as frequently occurring in jet fuels.[18,19] *H. resinae* was also found as a contaminant causing deterioration in fuel oils for marine and terrestrial turbine engines.[20] Ships with seawater displacement fuel tanks in which seawater is run in under the fuel to replace that consumed to maintain ship's center of gravity could be particularly susceptible to fungal attack.[21] In this respect, sulfate-reducing bacteria (SRB) usually found in natural water, including seawater, can also thrive in ships' storage tanks if an environment protected from air is provided. This task is usually accomplished by other aerobic bacteria in the water bottoms of jet and diesel fuels providing organic nutrients and a suitable low-oxygen concentration. The sludge formed at the fuel/water interface and on tank surfaces can behave like a separate, almost isolated phase in which conditions may be considerably different from those in the bulk of the fuel or water.[19]

The 2024 aluminum alloy samples exposed during two years to an *H. resinae* culture in a JP1/simplified mineral medium system showed a heavy mycelial growth forming dark patches on the metal surface (Figure 12.1, p. 199). A similar development of *H. resinae* on the walls of fuel tanks can result in penetration and breakdown of fuel tank coatings and corrosion of the metal. This kind of growth, which forms a blanket of dark brown slime at the water phase and oil/water interface, is the main cause of fuel pump, filter, and coalescer malfunction due to clogging blockages in jet aircraft. When fungal deposits were partially removed, it could be seen that they were associated with localized areas of attack on the metal surface (Figures 12.2 and 12.3, p. 199), mainly by pitting.

Yeast contamination (mainly *Candida sp.*) in seawater/diesel fuel systems encourage and facilitate *H. resinae* growth through pH reduction and extracellular biosurfactants production in restricted areas.[21] Microbial contamination of kerosene is more detrimental than contamination of gasoline. Due to the high rate of consumption of kerosene by jet engines, even small amounts of microbial sludge in the fuel become hazardous. The combination of microorganisms and their associated water creates major problems of fuel contamination, including fuel system filter plugging, fuel gauge malfunctions and fuel tank corrosion.[22]

During the past decade, the always-increasing activities related with oil production and subsequent processing, transport and storage have been frequently affected by MIC problems.[23] The corrosion and subsequent leakage of hydrocarbon fuel storage tanks caused important economic losses as well as different troubles related with oil and underground water contamination.[24,25] The relevant role played by fuel microbial contaminants in these undesirable effects is also increasingly recognized.[26,27] The corrosive attack is generally located at the bottom of the tanks where there is an active microbial population associated with free water. Electrochemical polarization techniques and current transients of aluminum, 2024 alloy, and mild steel were used in the presence of sterile and contaminated fuel to evaluate the role of the microorganisms in the corrosion process. Additionally, scanning electron microscopy (SEM) observations were made complemented with energy dispersive x-ray analysis (EDAX). Gas chromatography was used to follow the attack of hydrocarbon chains by microbial action.

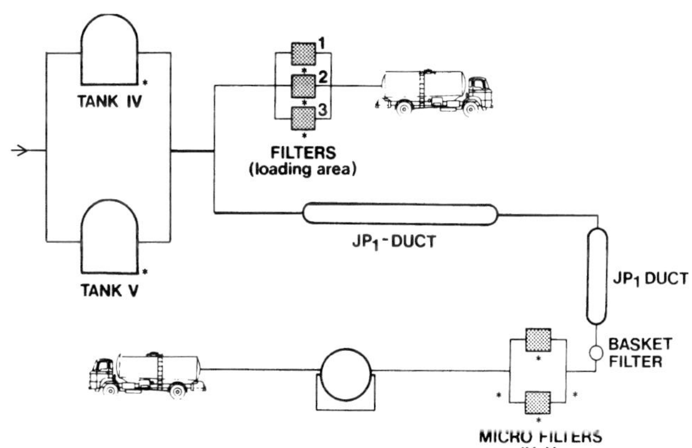

FIGURE 12.4
Simplified scheme of the fuel storage and distribution system.

Experimental

The strains of microorganisms used in the present work were isolated from a JP1 fuel storage and distribution system. A simplified scheme of this system is shown in Figure 12.4. Sampling areas are marked in the figure.

Conventional culture media for the isolation of fungi (Sabouraud agar) and bacteria (nutrient agar) were used. Identification of the fungal strains was made by phase contract microscopy. This type of observation was also employed for bacteria but complemented with gram staining and biochemical tests. After isolation and selection of microbial colonies, fungal and bacterial strains were transferred to a simplified mineral medium of the following composition: $CaCl_2$ 0.02 g; $MgSO_4 \cdot 7H_2O$ 0.42 g; $(NH_4)_2SO_4$ 1.00 g dissolved in distilled water to one liter. The pH was adjusted to 7.0 by addition of NaOH 0.1 N. When nitrate was used as an alternative nitrogen source for *H. resinae* growth, ammonium sulfate was replaced by potassium nitrate at a similar concentration. These media were also used as electrolytes for the electrochemical experiments.

Fungal and bacterial growth were made in Erlenmeyer flasks of 500 mL containing 50 mL of jet fuel as carbon source and 250 mL of simplified mineral medium. The inorganic components of the medium were sterilized in an autoclave while the fuel was sterilized by filtration through a membrane with 0.2 μm pore diameter.

Static conditions were chosen for growth, and temperature for incubation was 30°C (86°F).

The inocula for fungi were prepared by suspending spores taken from a Sabouraud agar slant (ten days old) in 2 mL of sterile medium. In the case of bacteria, the growth in a nutrient agar slant was transferred into 2 mL of sterile medium. The purity of fungal cultures was verified from time to time by microscopic observation of the cell morphology through phase contract microscopy.

Gram staining and a simplified scheme of biochemical tests were used for bacteria.

The water phase was transferred to the electrochemical cell after the stationary phase of growth was reached. Growth rate was followed by the plate dilution method for bacteria and by dry weight determinations of samples representing the whole volume of the aqueous phase for fungi.

The different species of fungi and bacteria isolated from sampling areas (Figure 12.4) are summarized in Table 12.1.

TABLE 12.1
Fungi and Bacteria Isolated from a Fuel Storage and Distribution System

Sampling Area	Total Number of Fungal Colonies*	Fungi	Partial Number of Fungal Colonies	Bacteria[(1)]
Tanks IV and V	1.1×10^5	*Hormoconis resinae*	4.75×10^4 +	*Pseudomonas sp.* +
		Aspergillus niger	0.50×10^4 -	
		Fusarium sp.	1.75×10^4 +	
		Aspergillus terreus	0.25×10^4 +	
Filters 1, 2 and 3 (loading area)	5.0×10^4	*Penicillium sp.*	1.00×10^4 -	
		Phoma sp.	0.255×10^4 +	
Before microfilters IV and V	4.0×10^4	*Aspergillus niger*	0.50×10^4 -	
		Trichosporon sp.	0.50×10^4 +	
		Phaecylomices sp.	1.00×10^4 -	
		Penicillium sp.	0.50×10^4 -	
Microfilters IV and V	7.5×10^5	*Hormoconis resinae*	0.50×10^4 +	*Serratia marcescens* +
		Phaecylomices sp.	5.50×10^4 -	
		Aspergillus terreus	0.25×10^4 +	
After microfilters IV and V	2.5×10^4	*Phaecylomices sp.*	0.50×10^4 -	
		Aspergillus niger		

*CFU/1
[(1)]Able to metabolize JP1 fuel

The electrochemical cell used has been previously described.[3] Working electrodes correspond to disks of pure aluminum (99.9%) and 2024 alloy or mild steel embedded in an epoxy resin and offering an effective area of 0.4 cm^2. Prior to each test, the electrodes were polished with several emery papers (200, 600 and 800) and finally with alumina paste (grain size 0.3 μm). Potentials were referred to a standard calomel electrode (SCE) through a Luggin capillary to minimize the ohmic drop. A platinum wire served as counter electrode. For potential dynamic curves, the sweep rate was 20 mV/min starting from -0.8V in the anodic direction and ending at the potential inflection corresponding to the pitting potential (Ep) of the sample. To more precisely define Ep values, current transients were occasionally made. SEM observations were made in an appropriate microscope. Occasionally, EDAX analysis of the metal samples was made complementing SEM observations. Gas chromatography analysis of fuel samples previously incubated with each microbial contaminant was made using sterile fuel samples for comparison.

Results and Discussion

Uptake of nitrate by H. resinae

The breakdown of passivating oxide films by aggressive anions such as chloride frequently causes corrosion of aluminum and its alloys in aqueous environments.[28] This process leads to severe pitting on the metal. It has been reported that water accompanying aircraft fuel contains chloride anions levels in the 10^{-2} M to 10^{-4} M concentration range.[29] This range of concentrations is similar to that of mineral media usually employed as electrolyte and culture medium in several publications on the MIC of aluminum alloys in fuel/water systems.[30]

An important decrease of Ep values for 2024 alloy with the chloride concentration in the 10^{-2} M to 10^{-4} M concentration range has been reported.[7] Conversely, phosphate and nitrate anions also present in mineral culture media compositions and in bilge water accompanying the fuel have an opposite effect. Phosphate anions, by their buffering action, could prevent the localized acidification necessary to initiate the pitting process.[31] Besides, nitrate anions are considered the most efficient pitting inhibitors of the corrosion of aluminum in chloride solutions. This fact explains the anodic values of Ep for aluminum alloys usually reported in mineral media (i.e., Bushness and Haas) containing nitrate.[7]

The growth of microbial contaminants of hydrocarbon fuels is conditioned by the basic elements C, O, H, N, and P as well as some others needed in much smaller amounts but nevertheless essential for a normal metabolism.[15] The percentage values of the main elements in cell composition on a dry weight basis (after Hopton) are presented in Table 12.2.[32]

TABLE 12.2
Main Elements in Cell Composition (Dry Weight Basis) (After Hopton[32])

> 1%	1% > 0.05%	0.05%
C (50)	S	B
O (20)	Cl	Mn
H (10)	K	Zn
	Na	Si
N (10)	Ca	Co
	Mg	I
	Fe	
P (3)	Cu	F

Figures in brackets are typical percentage values for dry bacterial cells.

For growth to occur, the environment must provide these elements in a form that can be assimilated as molecules. Carbon and hydrogen are abundantly supplied by the hydrocarbon molecules of the fuel (usually 95% of the product), although for these molecules an essential nutrient for their biodegradation is oxygen.[33] The rate of the oxidation of the hydrocarbon is oxygen-dependent. Low concentrations of oxygen in a stagnant tank result in low rates while high rates can be expected in active tanks where oxygen is carried into the tank when fresh fuel is added. Thus, microbial growth and the subsequent rate of hydrocarbon biodegradation would be mainly limited by the available concentration of nitrogen and phosphorus in the medium. The bulk of these elements is supplied by the hydrocarbon phase in the case of crude oil and in semi-processed material. However, nitrogen and phosphorus availability in kerosene is often the limiting factor to growth. These elements are generally present as nitrates and phosphates, either in the composition of the water phase or in the formula of mineral culture media used in laboratory experiments. It has been reported that in aircraft aluminum integral fuel tanks, although shortage of nitrogen restricts growth, it actually appears to enhance corrosion due to an enhanced production of extracellular organic acids.[34]

The incorporation of nitrate and phosphate into the biomass diminishes the level of corrosion inhibitors in the medium increasing the chloride/inhibitor ratio. The existence of a critical chloride/nitrate ratio (c.a. 1.66) at which catastrophic corrosion of 2024 and 7075 aluminum alloys (A97075) occurs has been reported.[35,36]

Previously reported information concerning nitrogen metabolism of *H. resinae* indicates that this fungus prefers the nitrate ion to the ammonium ion, and it seems it can preferentially select the former if there is a choice.[37,38]

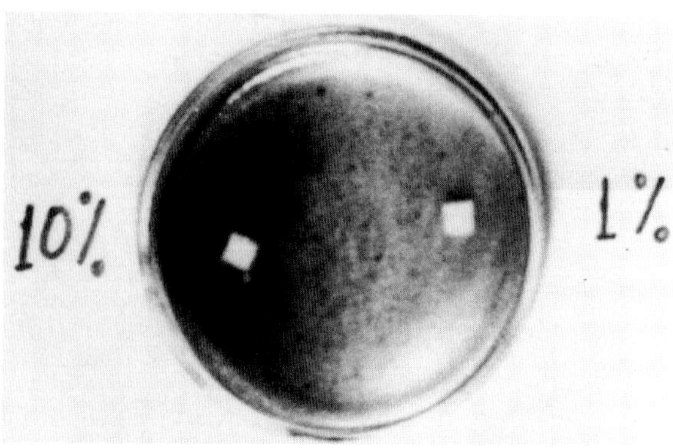

FIGURE 12.5
Photograph of the *H. resinae* auxanogram for nitrate assimilation. Potassium nitrate concentra-tions tested were 1% and 10% as indicated in the figure.

The auxanogram of *H. resinae* for nitrate assimilation using Lodder technique[39] with potassium nitrate as the only nitrogen source is shown in Figure 12.5. A marked increase in the fungal growth can be seen for the higher concentration (10%) of nitrate assayed.

The relationship between nitrate consumption and *H. resinae* growth (expressed as dry weight of mycelium) with time is shown in Figure 12.6. A

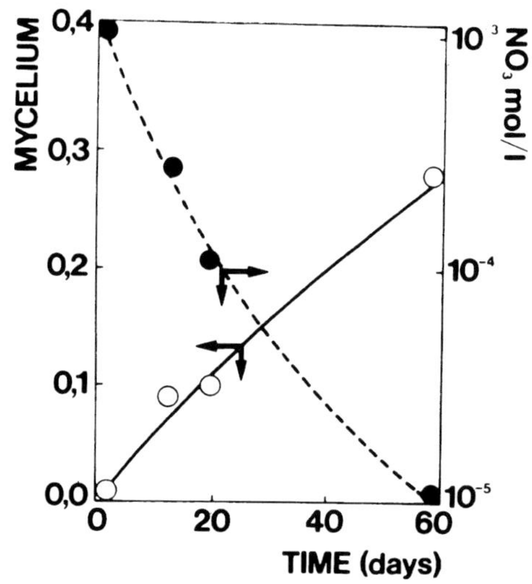

FIGURE 12.6
Relationship between nitrate uptake by *H. resinae* and fungal growth (expressed as mycelium weight) vs time.

variation of nitrate concentration between 10^{-3} M and 10^{-5} M is produced as a consequence of the fungal growth (Figure 12.6). The variation of the Ep values corresponding to 2024 aluminum alloy in sterile simplified mineral medium (with KNO_3) as a function of the chloride/nitrate ratio is shown in Figure 12.7. Ep values depend markedly on the chloride/nitrate ratio showing Ep variations in the order of 0.5 V for a two units of logarithm range.

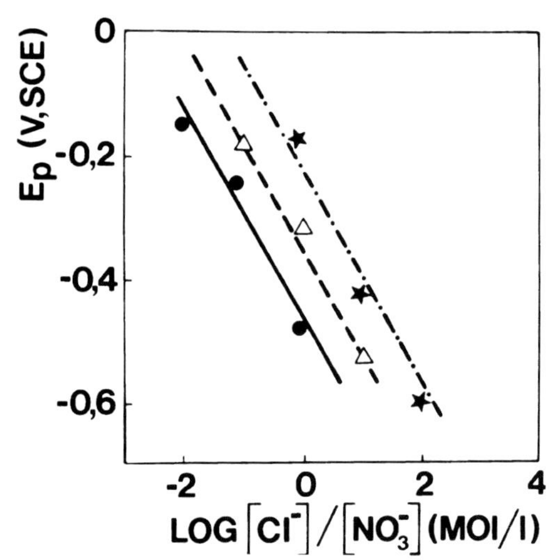

FIGURE 12.7
Variation of the Ep of 2024 aluminum alloy vs chloride/nitrate ratio in simplified mineral medium for different potassium nitrate concentrations; 10^{-2} M/L (-•-) ; 10^{-3} M/L (----); 10^{-4} M/L (—).

On the other hand, an important decrease in nitrate concentration has been reported (from 10^{-3} M to 10^{-5} M) for *Pseudomonas aeruginosa* cultures during the first 20 days of incubation in a mineral medium containing 1.3 x 10^{-3} M chloride anions.[40] In this case, an increase from 0.1 to 13 in the chloride/nitrate ratio could be expected. Abrupt decreases of the Ep for aluminum (in the order of 2V) have been reported when the chloride/nitrate ratio was increased from 0.1 to 1.0.[41] Thus, in aqueous media (such as bilge waters) containing chloride and nitrate anions the microbial uptake of nitrate is able to produce a marked decrease in the Ep of aluminum alloys.

Smaller effects than those produced by nitrate uptake can be expected as a consequence of phosphate anion consumption by microorganisms. The Ep variation of 2024 aluminum alloy with time of incubation of *H. resinae* cultures in simplified mineral medium using either ammonium sulfate or potassium nitrate as nitrogen sources is shown in Table 12.3. The more cathodic values found in ammonium containing medium is mainly due to the absence of corrosion inhibitors in the electrolyte. Lower pH values are reached in *H. resinae* cultures using an ammonium salt as nitrogen source when compared with those media containing KNO_3.

TABLE 12.3

2024 Alloy Ep Variation with Time of Incubation in *H. resinae* Cultures, Using Different Nitrogen Sources

Incubation Time (Days)	Ep (V, SCE) Nitrogen Source	
	$(NH_4)_2SO_4$	KNO_3
15	-0.210	-0.180
30	-0.380	-0.220
100	-0.370	-0.240

The morphology of the pitting attack induced by anodic polarization at the Ep of 2024 alloy in *H. resinae* cultures (with ammonium sulfate) revealed a crystallographic appearance like that reported for aluminum for chloride solutions (Figure 12.8).[42] Conversely, the pit pattern observed in *H. resinae* cultures employing KNO_3 as nitrogen source show a varied morphology of pits ranging from irregular to round-shaped pits (Figure 12.9). This pit pattern is similar to that reported in the literature for the pitting of aluminum in nitrate-containing solutions.[42] EDAX analysis of the attacked areas revealed a marked increase in the copper content and depletion in magnesium in agreement with previously reported results for aluminum corrosion induced by *H. resinae* in ammonium ion containing solutions.[9]

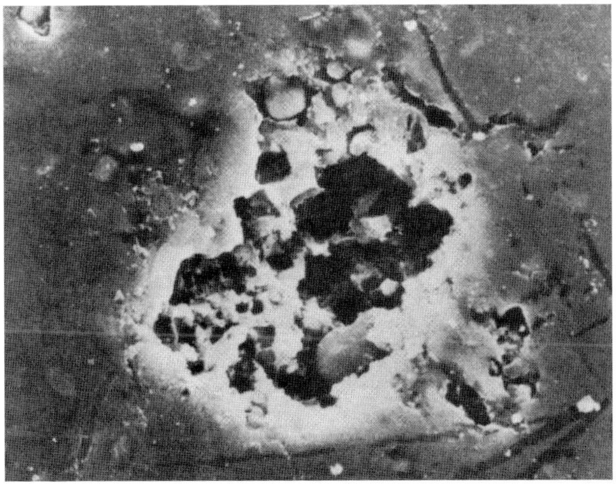

FIGURE 12.8
SEM microphotograph of the pitting attack on the 2024 aluminum alloy after anodic polarization over the Ep value in an *H. resinae* culture containing ammonium sulfate as nitrogen source. (Magnification: 1400X)

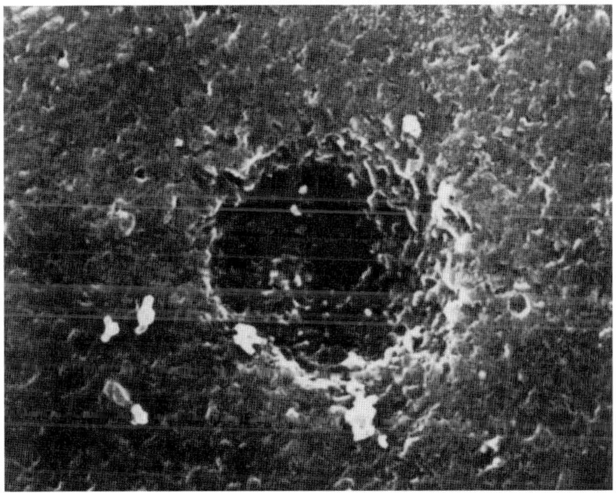

FIGURE 12.9
SEM microphotograph of the pitting attack on the 2024 aluminum after anodic polarization over the Ep value in an *H. resinae* culture containing potassium nitrate as nitrogen source. (Magnification: 5000X)

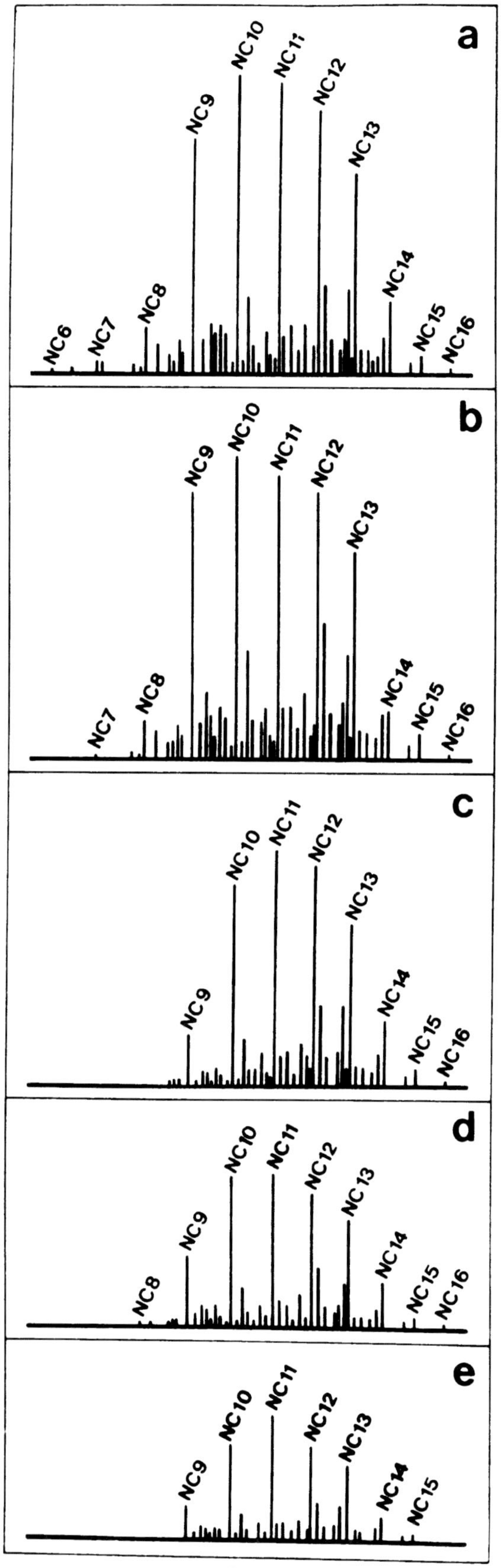

The chromatographic analysis of the fuel phase in contact with *H. resinae* cultures shows a preferential assimilation of n-alkanes of short chain-length which are removed first and then progressively longer n-alkanes. Besides, *H. resinae* cultures in KNO_3 containing media revealed a higher degree of degradation of the hydrocarbon chains (Figure 12.10).[15] This experimental fact also could account for the enhanced corrosion attack observed in these cultures.[37] It has been reported that the metabolism of n-alkanes in order of increasing molecular size accords with the lag phase phenomena of filamentous fungi growing on n-alkane mixtures.[43] During growth, all species produce surface-active agents (mainly phospholipid and fatty acids). The effect of these surface-active compounds in destabilizing the passivating film on aluminum alloys surfaces has been reported.[6] Long lag phases are observed for filamentous fungi growth on hydrocarbons, e.g. 85 days for *H. resinae*.[43]

The differential effect of fungal metabolism due to the nitrogen source used on the corrosive behavior of aluminum and its alloys can be extended to several species of fungi other than *H. resinae*. Different strains of *Aspergillus*, *Fusarium*, *Penicillium* and *Trichosporon* isolated from the fuel storage and distribution system were able to metabolize either ammonium or nitrate as nitrogen sources.

Preliminary results obtained with these species revealed that their effects on corrosion would be closely related to their degradation capacity of the aliphatic fraction of the fuel. Thus, for the different fungi tested their corrosiveness on aluminum increases according to the following se-

FIGURE 12.10
Degradation of the aliphatic fraction of JP1 fuel by *H. resinae* as shown by the chromatographs for different culture conditions. (a) JP1 blank (not incubated); (b) JP1 blank (incubated during 10 months in sterile conditions); (c) JP1 incubated during 10 months with an *H. resinae* culture (simplified mineral medium without nitrogen source); (d) JP1 incubated during 10 months with an *H. resinae* culture (simplified mineral medium with ammonium sulfate as nitrogen source); (e) JP1 incubated during 10 months with an *H. resinae* culture (simplified mineral medium with potassium nitrate as nitrogen source).

quence: *Penicillium sp.*, *Trichosporon sp.*, *Fusarium sp.*, *Aspergillus sp.*, *H. resinae*.

This sequence is similar to the one recently reported for the fuel degradation rate by several of the fungal species assayed in this work.[43]

Protective Action of Serratia Marcescens

There are few references in the literature on the action of kerosene-contaminating bacteria in relation to corrosion of fuel storage and distribution systems. In a previous publication, a different corrosive effect between two different strains of *Pseudomonas* isolated from ship fuel storage tanks was found.[5] By using electrochemical methods, it was determined that the corrosion ability was related to the acidification of the medium during growth. In this work, the two different bacterial species isolated from the fuel storage and distribution system (Figure 12.4) were used to assess their effect on the corrosion behavior of aluminum and 2024 aluminum alloys.

Several *Pseudomonas* species are referred to in the literature as common contaminants of petroleum fluids and derivatives (aromatic and paraffinic hydrocarbons, kerosene fuels, cutting oil emulsions) as well as asphalt and asphalt-based coatings.[13] Generally, their presence produces an alteration of fuel quality, leading to slime formation and blockage of filters and injection systems.[44] Several types of corrosion processes have been reported, mainly involving pitting corrosion.[45] However, the presence of *S. marcescens* has only occasionally been reported in association with secondary oil recovery systems, but no mention is made in the literature about its effect on corrosion.[13]

The Ep values obtained for aluminum and 2024 alloy in *Pseudomonas sp.* and *S. marcescens* cultures of the same age are shown in Table 12.4 Ep values of the same metals in sterile medium. A marked increase of Ep values of aluminum and 2024 alloy towards anodic potentials is observed in the presence of *S. marcescens* or in sterile medium showing a passivating effect of these bacteria on the metal surface. The importance of local acidification, many times enhanced by adhesion processes taking place at the metal/mycelium interface, accounts for some of the specific

TABLE 12.4

Ep Values for Aluminum and 2024 Aluminum Alloy in Cultures of *Pseudomonas sp.*, *S. Marcescens* and Sterile Simplified Mineral Medium (Time of Incubation: 15 Days)

Microorganism	Metal	pH Water Phase	Ep V, SCE
S. marcescens	Aluminum 2024 alloy	6.50	0.42 -0.10
Pseudomonas sp.	Aluminum 2024 alloy	5.75	-0.29 -0.56
Blank (sterile medium)	Aluminum 2024 alloy	7.00	0.01 -0.37

effects of microbial contaminants (*H. resinae*) on jet fuels in the corrosion of aluminum alloys in fuel/water systems.[46] Acidity can prevent repassivation, hindering the protective oxide film formation.

Thus, in acidic conditions, pitting of the metal due to chloride anions occurs at more cathodic potentials than in neutral solutions. The acidity, however, is not the only factor accounting for the aggressiveness increase observed in the presence of microbial contaminants.

The important effect of chloride ions in the passivity breakdown is generally enhanced by the metabolic products of microbial hydrocarbon degradation, mainly mono-, di-, and tricarboxilic acids, although the corrosiveness of these compounds can be reversed to a passivating effect when neutral pH values are reached in the aqueous phase.[47] These neutral values are more frequently found in bacterial cultures in which low pH values of the aqueous environment are required to allow microbial growth.[19]

Microbial growth induces complex modifications of the system through oxygen consumption, production of metabolites, and cellular lysis products. The first effect leads to a decrease in the cathodic rate favoring the protection of the metal, but the second (production of acidic organic substances) counteracts the oxygen depletion, producing a net increment in the cathodic rate. Thus, the amount of hydrocarbon degraded by microorganisms would be an important factor in explaining bacterial activity in relation to corrosion.

The different conversion percentages of n-C8 to n-C15 alkanes after incubation for 15 days with *Pseudomonas sp.* or *S. marcescens* are summarized in Table 12.5. Sterile control values have been taken into account. A higher degree of hydrocarbon conversion was found with *S. marcescens* cultures. With both microorganisms, n-C8 to n-C10 alkanes are mainly degraded in concordance with the degradation scheme presented by fungi for short times of incubation.[43] Although it has a slower degrading capacity, lower pH values were reached in the presence of *Pseudomonas sp.*, accounting for its higher aggressiveness to the metal. Therefore, at near neutral pH values, such as those encountered with *S. marcescens* cultures, organic metabolites will be present as anions. Their effect will be to passivate the aluminum surface through the production of several types of insoluble precipitates on the attacked areas, as has been previously reported.[47,48]

TABLE 12.5

Percentage of Conversion of n-C8 to n-C10 Alkanes After 15 Days of JP1 Incubation with *Pseudomonas sp.* and *S. Marcescens* Cultures

Microorganism	% Conversion
S. marcescens	5.22
Pseudomonas sp.	1.31

Conversion of n-C11 to n-C15 alkanes was nil during this time of incubation.

Despite the results reported here, in practical situations the protective action of *S. marcescens* can be reversed through the modification of restricted areas of the environment at the metal/solution interface, mainly through the presence of microbial consortia in the biofilms or slime deposits formed at the walls of the fuel storage tanks. SEM observation of several metal samples after anodic polarization over Ep values shows a crystallographic pattern of the pits similar to the one reported for aluminum pitting in chloride anions in aqueous media (Figures 12.11 and 12.12). As microbial metabolites seem not to modify pit morphology, it is possible to assign the main corrosive effect to chloride anions assisted by low pH values and organic acidic metabolites. The action of these metabolites, present as anions at neutral pH, and oxygen depletion due to microbial growth are the two main effects accounting for metal protection.

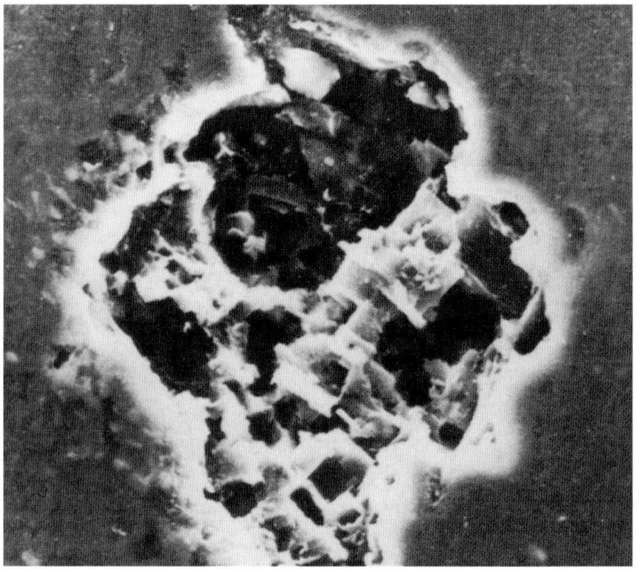

FIGURE 12.11
SEM microphotograph of the pitting attack on the aluminum after anodic polarization over the Ep value in a *S. marcescens* culture. (Magnification: 1200X)

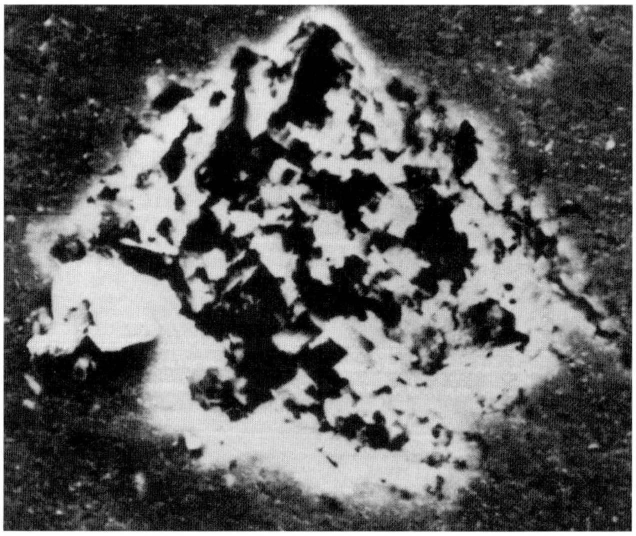

FIGURE 12.12
SEM microphotograph of the pitting attack on the aluminum after anodic polarization over the Ep value in a *Pseudomonas sp.* culture. (Magnification: 1200X)

Importance of Localized Acidification in the Corrosive Action of H. Resinae on Mild Steel

There are few references on the effect of fungal contaminants of kerosene fuels on the corrosion of mild steel, although this is one of the most commonly used materials employed in the construction of fuel storage and distribution systems. In the case of ship storage tanks ballasted with seawater, the inorganic composition of the water phase is characterized by high chloride levels due to the seawater contribution (in the order of 0.5 M). The anodic polarization curve of mild steel in artificial seawater shows three main regions. The first one, between -0.72 V and -0.66 V, corresponds to an active dissolution of the metal. At potentials more positive than -0.66 V the active/passive transition occurs followed by a passive region, which extends up to -0.58 V. If the potential value exceeds -0.58 V, a marked current increase is observed and pits grow on the metal surface. The anodic potentiodynamic polarization curves in sterile seawater at its usual pH (7.8) and in sterile seawater plus two different amounts of a *H. resinae* culture (stationary phase of growth) with a pH adjustment of the medium to pH 7.8 with N NaOH show two distinct effects:

- The addition of *H. resinae* metabolites to seawater at pH 7.8 shows an increase of current with increasing metabolite concentrations. No change of Ep values is noticed, and the three regions (active, passive, pitting) of the polarization curve remain unaltered.
- When the addition of *H. resinae* cultures to seawater is made without pH adjustment, a dramatic change of the polarization curve is observed. In this case, a final pH of 4.1 is reached for the solution. No passivity is found, and higher current densities are obtained for the mild steel. To compare the effect of one of the main metabolites of hydrocarbon degradation by *H. resinae* with the fungal cultures, a solution of citric acid in sterile seawater is used at a similar pH value. A similar electrochemical behavior of the steel is obtained in this case (Figures 12.13 and 12.14).

These experiments could be considered approaches to two real situations encountered in service. The first (Figure 12.13) would correspond

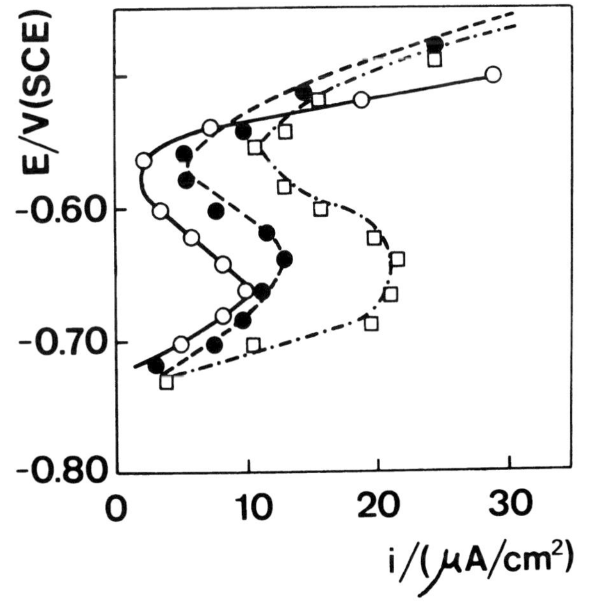

FIGURE 12.13
Anodic polarization curves of mild steel in (o) artificial seawater (pH 7.80), (•) artificial seawater + 5 mL of an *H. resinae* culture (pH 2.9) adjusted to 7.8 by addition of N NaOH, (□) artificial seawater + 25 mL of an *H. resinae* culture (pH 2.9) adjusted to 7.8 by addition of N NaOH.

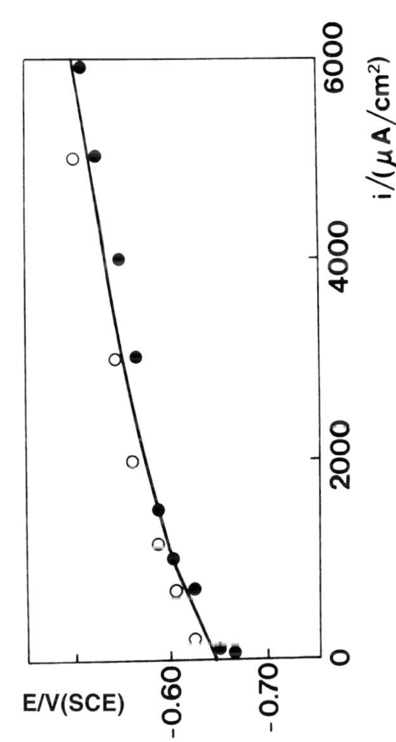

FIGURE 12.14
Anodic polarization curves of mild steel in (o) artificial seawater + 25 mL of an *H. resinae* culture (pH 4.1) and (•) artificial seawater + citric acid (pH 4.0).

to the metal contact with water bulk where pH changes are not expected due to the buffer capacity of seawater. The other case (Figure 12.14) would correspond to the metal regions where microbial colonies were attached. In this situation, a local drop of pH can be expected due to the organic acid production by the fungus. Localized acidification results in a dissolution of the protective oxide film with the subsequent pitting of the bare metal. In all cases, an increment of the aggressive metabolites concentration can be considered in the neighborhood of the colonies attachment.

These considerations also are valid in explaining the preferential pitting of aluminum alloys in the region where colonies are present. The properties of the water phase of the sludge at the tank bottoms are not likely to be the same as the bulk of the water on fuel in a tank. The sludge may thus be considered a microenvironment where microorganisms can generate chemical changes not characteristic of the tank contents as a whole. The previous electrochemical experiments clearly illustrate the marked differences that can arise in the electrochemical behavior of the metal in both situations.

It was recently reported that the fixation points of the mycelium to the metal act as nucleation centers for pitting, which shows the importance of the fungal adhesion processes in enhancing metal dissolution at those restricted areas.[30]

A similar effect due to local acidification on the corrosive behavior of mild steel was obtained for other fungal species isolated from the fuel storage and distribution system, such as *Aspergillus sp.*, *Penicillium sp.*, and *Fusarium sp.*

Microbial Contaminants of Kerosene Fuels in Corrosion

As an idealization of the different effects on the corrosion of fuel storage and distribution systems due to the microbial contaminants of kerosene fuels, a sequence like that shown in Figures 12.15 to 12.18 can be suggested.

In Figure 12.15, an active microbial growth takes place in the water phase. The fuel is used to supply the main nutrients for microbial growth. As a consequence of hydrocarbon degradation, organic acidic compounds are poured into the medium. An oxygen gradient probably is established due to microbial respiration. In the water phase, several inorganic anions (Cl-, $SO_4^=$, NO_3-) usually are present, affecting the electrochemical behavior of the metal. The action of metabolic products would be highly conditioned by the pH values at the metal/solution interface. According to the scheme depicted in Figure 12.16, an active interchange of oxygen and proton is originated by microbial metabolism through the different interfaces formed on the metal. A complex interface structure derives from the presence of biofilm, microbial colonies and inorganic passive films

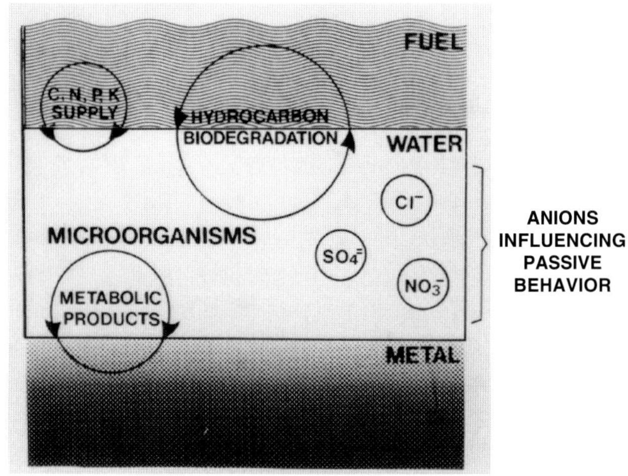

FIGURE 12.15
Simplified scheme of microbial growth effects in a fuel/water system at its initial stage.

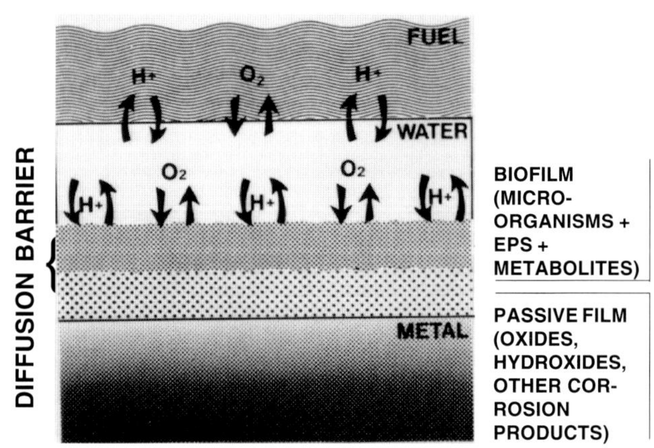

FIGURE 12.16
Simplified scheme of oxygen and proton interchange through the different organic and inorganic layers at the metal/solution interface.

(mainly oxides, hydroxides and corrosion products). Biofilm contains sessile bacteria and their extracellular polymeric substrates (EPS).

The existence of these two different components of the slime at the tanks is generally overlooked in the interpretation of MIC. EPS present in the biofilm usually leads to the formation and a diffusional barrier to the exchange of elements between the metal and the aqueous environment.[49] This is particularly hazardous when, due to low oxygen concentration and adequate nutritional supply, SRB grow in the inner layers of the biofilm, creating highly corrosive areas due to the high sulfide concentration. The development of the biofilm on the metal surface can lead to the formation of differential aeration cells.[50]

Figures 12.17 and 12.18 summarize the process of passivity breakdown and pitting initiation in localized areas of the metal surface. A synergistic action of organic acidic metabolites (mainly phase and fatty acids), low pH values and aggressive anions present in the water phase (mainly chloride) is produced in restricted areas of the metal/solution interface. The surfactant activity of fatty acids occasionally assisted by mechanical effects could account for the breakdown of passive film continuity at certain points of the metal surface. Once this breakdown is produced, the initiation and progression of pitting is established. The existence of microbial consortia in the vicinity may cause an enhanced corrosion due to the probable presence of biogenic sulfides derived from SRB metabolism.

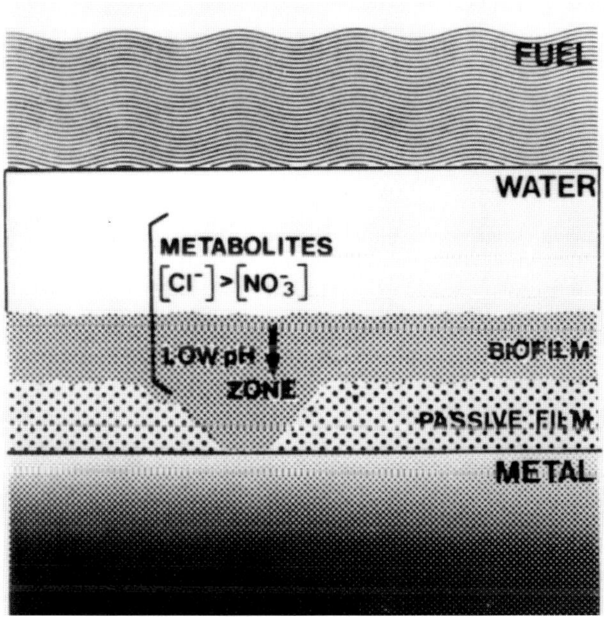

FIGURE 12.17
Simplified scheme of the initiation of passivity breakdown at the metal/solution interface.

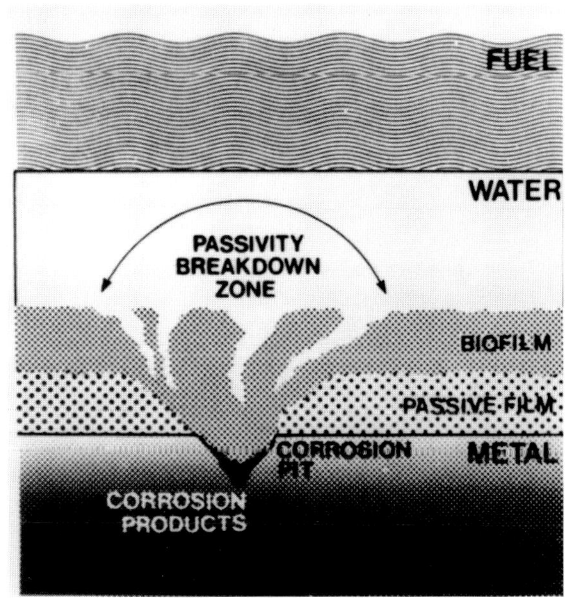

FIGURE 12.18
Simplified scheme of the initiation of pitting attack on the metal surface.

These complex sequences of interactions occurring at the metal/aqueous environment interface have to be properly evaluated when corrosion protection and prevention procedures are considered.

Acknowledgments

The authors would like to thank the technical assistance of Dr. Jeffrey Meyer and Mrs. Phyllis Chase (AMOCO Research Center, Naperville, Illinois) for gas chromatographic analysis of jet fuel samples. Aluminum alloys were kindly supplied by EMBRAER (Empresa Brasileira de Aeronautica S.A.). The laboratory assistance of Mr. Oscar Pardini is also gratefully acknowledged.

References

1. D.G. Parbery, *International Biodeterioration Bulletin* 4, 1(1968): p. 79.
2. P. McKenzie, A.S. Akbar, J.D. Miller, "Fungal Corrosion of Aircraft Fuel Tank Alloys," Tech paper, (London, UK: 1977), p. 37.
3. R.C. Salvarezza, H.A. Videla, *Anales Asociacion Quimica Argentina* 66, 3(1978): p. 317.
4. M.F.L. de Mele, R.C. Salvarezza, H.A. Videla, *International Biodeterioration Bulletin* 15, 2(1979): p. 39.
5. R.C. Salvarezza, M.F.L. de Mele, H.A. Videla, *International Biodeterioration Bulletin* 15, 4(1979): p. 125.
6. R.C. Salvarezza, M.F.L. de Mele, H.A. Videla, *British Corrosion Journal* 16, 3(1981): p. 162.
7. R.C. Salvarezza, M.F.L. de Mele, H.A. Videla, *Corrosion* 39, 1(1983): p. 26.
8. H.A. Videla, "Mechanisms of MIC," *Proceedings of Argentina—SA Workshop on Biodeterioration* (CONICET-NSF) (Sao Paulo, Brazil: Aquatec Quimica S.A., 1986), p. 43.
9. H.A. Videla, "The Action of *Cladorsporium resinae* Growth on the Electrochemical Behavior of Aluminum," *Biologically Induced Corrosion*, NACE-8 (Houston, TX: NACE, 1986), p. 215.
10. R.V. Miller, W.C. Herron, A.G. Krigrens, J.L. Cameron, B.M. Terry, *Materials Protection* 3, 9(1984): p. 60.
11. W.P. Iverson, *Electrochemical Technology* 5 (1967): p. 77.
12. H.G. Hedrick, *Materials Protection* 9, 1(1970): p. 27.
13. J.B. Davis, *Petroleum Microbiology* (Amsterdam, Holland: Elsevier Publishing Co., 1967), p. 502.
14. R.J. Watkinson, "Hydrocarbon Degradation," *Microbial Problems and Corrosion in Oil and Oil Products Storage* (London, UK: The Institute of Petroleum, 1984), p. 50.
15. W.B. Engel, F.E. Swatek, *Dev. Indust. Microbiol.* 7(1966): p. 351.
16. S. Holmes, "Microbiology of Hydrocarbon Fuels," *Proceedings of 2nd International Conference on Long-Term Storage Stabilities of Liquid Fuels* (San Antonio, TX: Southwest Research Institute 1986), p. 336.
17. P.F. Cannon, *Microbiological Sciences* 3, 6(1986): p. 169.
18. N. Bruce, *International Biodeterioration Bulletin* 18, 3(1982): p. 87.
19. R.A. Neihof, M. May, *International Biodeterioration Bulletin* 19, 2(1983): p. 59.
20. R.N. Smith, B. Crook, "Biocides in Fuels," *Biocides in the Oil Industry* (London, UK: The Institute of Petroleum, 1982), p. 69.
21. R.A. Neihof, "Microbes in Fuel: An Overview with a Naval Perspective," *Proc. 2nd Intern. Conf. on Long-Term Storage Stabilities of Liquid Fuels* (San Antonio, TX: Southwest Research Institute, 1986), p. 215.
22. C. Genner, E.C. Hill, "Fuels and Oils," *Microbial Biodeterioration* (London, UK: Academic Press, 1981) p. 259.
23. R.G. Edyvean, L.A. Terry, G.B. Picken, *International Biodeterioration* 21, 4(1985): p. 277.
24. L. Schon, P. Atterby, *British Corrosion Journal* 8(1973): p. 38.
25. G.A.H. Roberts, *British Corrosion Journal* 4(1969): p. 318.
26. S. Maxwell, W.A. Hamilton, "The Activity of Sulphate-Reducing Bacteria on Metal Surfaces in an Oil Field Situation," *Biologically Induced Corrosion* NACE-8 (Houston, TX: NACE, 1986), p. 284.
27. E.C. Hill, J.L. Shennan, R.J. Watkinson, eds., *Microbial Problems in the Offshore Oil Industry* (Sussex, UK: John Wiley and Sons Ltd., 1987).
28. Z.A. Foroulis, J.J. Thubrikar, *Journal of the Electrochemical Society* 122(1975): p. 1,296.
29. J.A. Scott, E.C. Hill, "Microbial Aspects of Subsonic and Supersonic Aircraft," Microbiology Symposium (London, UK: The Institute of Petroleum, 1971), p. 27.
30. R.C. Salvarezza, H.A. Videla, "Electrochemical Behavior of Aluminum in *Cladorsporium resinae* Cultures," *Biodeterioration* 6 (Farnham Royal, Slough: CAB International and the Biodeterioration Society, 1986), p. 212.
31. J.R. Galvele, "Present State of Understanding of the Breakdown of Passivity and Repassivation," *Proc. of the 4th International Symposium on Passivity* (Airlie, VA: 1977), p. 285.
32. J.W. Hopton, "Influence of Physical and Nutritional Conditions on Microbial Growth," *Microbial Problems and Corrosion in Oil and Oil Products Storage* (London, UK: The Institute of Petroleum, 1984), p. 32.
33. R.A. King, J.F. Stott, "Corrosion Hazard Assessment," *Microbial Problems and Corrosion in Oil and Oil Products Storage* (London, UK: The Institute of Petroleum, 1984), p. 93.
34. C. Rivers, "The Growth, Metabolism and Corrosive Effects of *Cladorsporium resinae*," Ph.D. Thesis, Corrosion and Protection Centre, University of Manchester, 1973.
35. A.M. McKissick Jr., A.A. Adams, R.T. Foley, *Journal of the Electrochemical Society* 117 (1970): p. 1,459.

36. A.A. Adams, K.E. Eagle, R.T. Foley, *Journal of The Electrochemical Society* 119(1972): p. 1,692.
37. D.M. Mardsen, *Mycologia* 46(1954): p. 161.
38. D.G. Parbey, *Material und Organismen* 6, 3(1971): p. 161.
39. J. Lodder, N.S.W. Kreger, *The Yeasts. A Taxonomic Study* (Amsterdam, Holland: Van Rij North Publishing Co., 1952).
40. C.G. Blanchard, C.R. Goucher, *Developments in Industrial Microbiology* 7(1966): p. 342.
41. H. Heinrich, N. Kanani, H.G. Feller, *Aluminum* 54(1978): p. 1.
42. J.R. Galvele, S.M. de Demichelis, I.L. Muller, S.B. Wexler, I.L. Alanis, "Critical Potentials for Localized Corrosion of Aluminum Alloys," *Localized Corrosion* (Houston, TX: NACE, 1974), p. 580.
43. N.D. Lindley, J.F. Pedley, S.P. Kay, M.T. Heydeman, *International Biodeterioration* 22, 4(1986): p. 281.
44. E.C. Hill, "Biodegradation of Petroleum Products," *Petroleum Microbiology*, ed. R.M. Atlas (New York, NY: Macmillan Publishing Co., 1984), p. 579.
45. J.D.A. Miller, "Metals," *Microbial Biodeterioration* (London, UK: Academic Press, 1981), p. 165.
46. D.J. Hansen, D.J. Tighe-Ford, G.C. George, *International Biodeterioration Bulletin* 17, 4(1981): p. 103.
47. B.W. Samuels, K. Sotoudeh, R.T. Foley, *Corrosion* 37, 2(1981): p. 92.
48. H. Bohni, H.H. Uhlig, *Journal of the Electrochemical Society* 116(1959): p. 906.
49. W.G. Characklis, *Biotechnology and Bioengineering* 23(1981): p. 1,923.
50. J.W. Costerton, G.G. Geesey, P.A. Jones, "Bacterial Biofilms in Relation to Internal Corrosion Monitoring and Biocide Strategies," CORROSION//87, paper no. 54 (Houston, TX: NACE, 1987).

Pitting of Galvanized Steel in a Cooling Tower Basin

CORROSION/81 PAPER No. 130[2]

By R.A. Schultz

Two separate but identical cases of pitting of galvanized steel in a cooling tower basin due to bacteria are described. Galvanized steel is commonly used in small cooling towers associated with commercial air conditioning systems. These cases are unique because they involved no discrete mounds or deposits generally characteristic of bacteria corrosion and also because the organisms defied all attempts at chemical control.

In the first case, a two-cell, induced-draft cooling tower was being operated on an air conditioning system in Milwaukee for approximately five months per year. Makeup was filtered and chlorinated city water drawn from Lake Michigan.

When the tower was started up in 1975, chemical treatment consisted of a zinc, phosphonate, and organic type corrosion inhibitor added frequently to maintain a continuous residual, and a monochlorophenate compound for general biological control. The biocide was added approximately every one to two weeks, depending on whether algae were seen in the cooling tower. The pH was generally between 7.5 and 8.0.

The system operated without apparent problems for three years, was treated during the operating season, and drained over the winters. Other than the occasional algae that triggered biocide addition, no deposits or biofouling were observed during this period. Periodic inspections of the tower and related equipment indicated the corrosion inhibitors were doing their job. Virtually no debris was accumulating in the tower other than a small quantity of sand and similar particles, which are normal in such open systems.

In mid-July of the fourth season, an operator noticed what appeared to be "depressions" in the bottom of the cooling tower basin. Closer examination showed these to be almost perfectly hemispherical pits in the galvanized steel ranging in size up to approximately 2 cm (3/4 in.) in diameter. No discrete mounds or deposits were found, but within each pit was a very loose collection of fine dark brown particles. Figure 12.19 (p. 200) shows the appearance of pits with residue intact. Gentle agitation caused the particles to be washed away, leaving no visible deposit or corrosion product. Figure 12.20 (p. 200) shows typical pits after the basin was drained and dried. Careful observation showed that the pits were increasing in both number and size at a rapid rate so that significant differences could be seen over a period of 24 to 48 hours.

Microbiological analysis of the debris from the pits found high counts of both aerobic sulfur-oxidizing bacteria, *Thiobacillus,* and anaerobic sulfate reducers, *Desulfovibrio.* It was suspected that other strains were present as well, but they were not identified in this program. Several methods were tried to minimize corrosion by controlling bacteria. These included:

- Higher than normal doses of the monochlorophenate biocide,
- High doses of an organic amine/organotin biocide added every day for one week,
- Continuous chlorination with sodium hypochlorite to maintain a free chlorine residual of over 3 mg/L (an exceptionally high level) for one week, and

---

[2]Originally published in *Materials Performance* 20, 8(1981): p. 44.

- A high level of quaternary ammonium chloride biocide was added, and the pH adjusted upward over a four-hour period to approximately 12 by addition of sodium hydroxide. At the end of this four-hour period, the tower water was drained and the tower flushed with a mild acid to neutralize any remaining alkalinity.

None of these treatments proved effective. The pitting continued to grow during this period and penetrated the tower basin at several locations. As pinholes developed, metal screws were inserted to temporarily stop leaks. During this entire time, the tower water remained clear, with turbidity only in the pits. At no time was corrosion found in other parts of the cooling water system. Finally, at the end of the season, an epoxy coating was applied to all basin surfaces. Since then, the tower has operated for two seasons, using the original chemical treatment with no signs of further deterioration.

Other cooling towers, including some with galvanized steel basins, had operated in the same vicinity, for the same period, without any signs of a similar problem. It was concluded that this was a "freak" occurrence requiring a particular combination of bacteria that was unlikely to recur. We also concluded that the zinc surface layer was somehow important to the mechanism, perhaps acting as a source of energy for these particular bacteria.

The following summer, again in July, a second incident occurred. This tower also was in the Milwaukee area, but not in the general vicinity of the first case. This case was similar to the first except that a different corrective action was tried.

Since it appeared that the bacteria might use zinc somehow in their metabolism, the zinc containing corrosion inhibitor was replaced with a completely organic phosphonate-triazole-molybdate. The monochlorophenate biocide was also replaced with a newer brominated compound that was added daily in shock doses. Analyses showed high counts of aerobic bacteria before each biocide addition, dropping to almost zero one hour after addition of biocide, but a high count developing again within 24 hours. Feed of the biocide was then altered to maintain continuous residual in the tower. Analyses showed this was even less effective, giving continuous high aerobic counts.

Like the first case, this case finally was solved by epoxy coating the tower basin. An extensive literature search revealed no similar incidents. There is good reason to suspect, however, that this phenomenon can occur at other locations because these materials of construction, operating conditions, and types of bacteria are common.

Microbiological Corrosion Control in a Cooling Water System[3]

By D.G. Honneysett, W.D. van den Bergh, and P.F. O'Brien

The failure of a corrosion control program in a closed cooling water system coincided with the use of reclaimed sewage water and the contamination of the system with oil. Other problems were increased corrosion rates, downward pH excursions, increased fouling by corrosion byproducts, and increased microbiological activity in the system. The major cause of corrosion was microbiological in origin.

The unsuccessful use of a biocide led to the initiation of a full-scale microbiological investigation. The nature of the microflora was determined, biocide selection tests made, and an effective control treatment program initiated. Chromate corrosion treatment was replaced by a coordinated program using an organic filming corrosion inhibitor, a polyacrylate/phosphonate dispersant, and a combination of biocides.

Introduction

The mitigation of a corrosion problem developed in a closed cooling system at an integrated steel works illustrated by this work includes:

- Identification of the troublesome microflora and documentation that the corrosion was microbiologically induced,
- Establishment that conventional treatment procedures for control of corrosion caused by electrochemical mechanisms were ineffective under existing conditions in controlling MIC,
- Development of a treatment program that mitigated an existing MIC problem and avoided the aggressive cleanup procedures that increase the potential for corrosion, and
- Application of a coordinated maintenance treatment program that effectively prevented the reestablishment of MIC and also effectively maintained inhibition of corrosion caused by other mechanisms.

Review of Problem

The bloomcaster closed-loop cooling system is used to cool various sections of a continuous casting machine process in an integrated steel works located in Newcastle, South Africa (Figure 12.21).

The cooling system operated for several months with high efficiency; there were no observed operational problems. Microbiological populations were monitored weekly with total plate counts of the circulating cooling water (Table 12.6). Corrosion rates were routinely measured by using linear polarization electrode techniques and by measuring weight loss of corrosion coupons placed at various locations in the circulating water system. Corrosion rates were consistently less than 25 µm/y to ~1 mpy. Fouling was monitored by using test heat exchangers and determining heat flux across heat exchanger processes.

Makeup water for the system initially was softened well water containing a total hardness of less than 2 mg/L (as $CaCO_3$). The pH of the

[3]Originally published as CORROSION/85, paper no. 291 (Houston, TX: NACE, 1985)

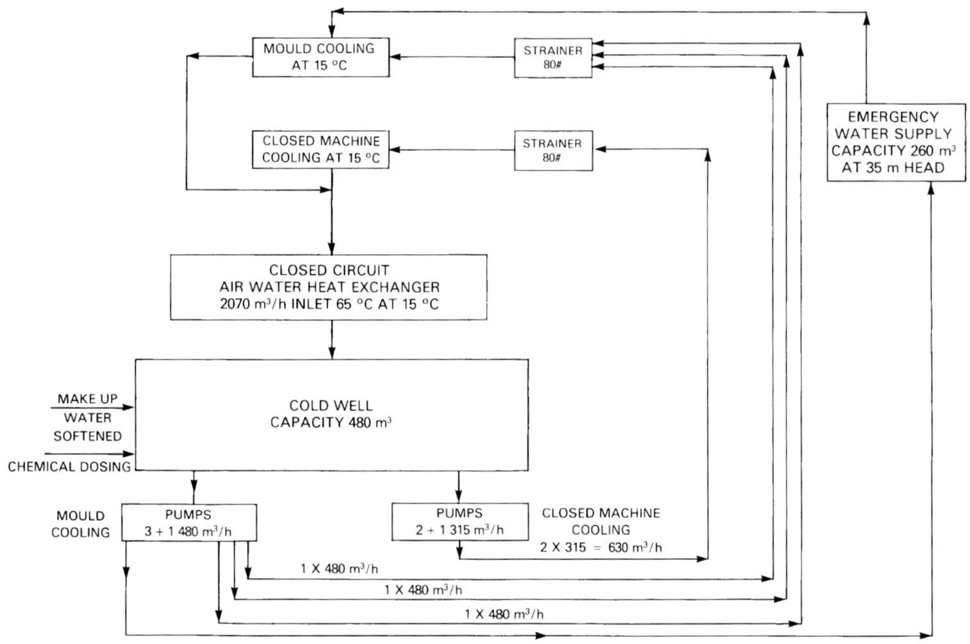

FIGURE 12.21
Flow diagram, bloomcaster closed-loop cooling water circuit.

makeup water was approximately 8.0. Other characteristics are shown in Table 12.7.

A low chromate zinc plus phosphonate treatment program was used to control corrosion and scale/fouling in the circulating cooling water system. Chromate concentration of 10 mg/L (as CrO_4) was maintained in the circulation water. A microorganism control program had not been used routinely because the total plate count data indicated the microflora population was relatively low, i.e., less than 50,000 colonies per mL (Table 12.6, April to October 1983). When the total count exceeded this level, a slug dosage of an organosulfur biocide was made to provide a concentration of 50 mg/L, based on the volume of water in the system.

The first indication of the problem was a steady increase in the corrosion rate. Measurements changed from less than 25 μm/y to more than 75 μ/y over four to five weeks.

Concurrent with the increased corrosion rates was a dramatic increase in the microbiological populations. The total plate count increased from fewer than 50,000 colonies per mL aerobic bacteria to greater than 1,500,000 colonies per mL. The occurrence of anaerobic sulfate-reducing bacteria (SRB) began to increase approximately two to three months later. Therefore, a microbiocidal treatment program was begun. Slug doses of the organosulfur biocide at 50 mg/L were made to the circulating water system three times per week. The microbiological population, however, never

TABLE 12.6

Monthly Averages of Microbiological Survey Data for the Periods Before, During, and After the Mitigation of the MIC Problem

	Total Counts - Colonies per mL	
Date	**Aerobic Bacteria**	**Anaerobic SRB**
Apr 1983	5.0×10^5	0
May 1983	8.2×10^5	0
Jun 1983	8.3×10^4	20
July 1983	8.6×10^5	0
Aug 1983	4.9×10^4	2
Sep 1983	6.0×10^5	0
Oct 1983	3.0×10^2	0
Nov 1983	3.4×10^6	0
Dec 1983	2.0×10^5	0
Jan 1984	6.0×10^6	0
Feb 1984	3.6×10^6	0
Mar 1984	9.4×10^6	0
Apr 1984	6.4×10^6	108
May 1984	6.5×10^6	123
Jun 1984	3.8×10^7	TNTC
Jul 1984	4.2×10^3	0
Aug 1984	1.0×10^4	0

TABLE 12.7

Typical Analysis of Makeup and Circulating Waters

Item	Unit	Circulating	Makeup
pH	pH	8.0	8.1
TH	mg/L $CaCO_3$	2.0	1.2
Cond	μS/cm	420.0	340.0
Ca	mg/L $CaCO_3$	1.2	0.8
M. Alk	mg/L $CaCO_3$	70.0	
SO_4	mg/L	95.0	80.0
Cl	mg/L	21.0	21.0
Suspended solids	mg/L	10.0	2.0
Oils	mg/L	4.0	1.0
Total PO_4	mg/L	14.0	2.0
Total Fe	mg/L	3.8	2.8

Note: Chemical characteristics did not change significantly with the inclusion of recovered waste water as part of the makeup water.

returned to its original low level (Table 12.6, November 1983 to June 1984). The weekly microbiological monitoring tests also indicated increasing numbers of SRB from March to June 1984.

The increase in microbiological population can be explained by the fact that treated industrial waste water was used as part of the makeup water source for the system. The commingled well water and recovered waste water was pretreated by softening prior to use as makeup water for the system. The chemical characteristics of the pretreated commingled makeup water and the circulating water did not change significantly from those shown in Table 12.7. The use of waste water as part of the makeup was necessary because a severe drought had limited the supply of well water previously used for makeup.

To this point, no changes were made in the corrosion control treatment program despite the increased corrosion rates. After approximately two months, the heat transfer efficiency dropped dramatically, and production losses were incurred during periods of high ambient temperatures. The heat exchanger cells were inspected and found heavily fouled (Figures 12.22 and 12.23, p. 200). For the first time, oil contamination was noted. The heat exchangers were cleaned up with high-pressure water, and oil dispersant was added to control the oil contamination.

Investigations were also made to ascertain the cause of the oil contamination. The oil coolers had corroded through from the oil side. This problem was rectified by replacing the coolers.

An inspection one month later indicated the system had fouled again; however, the new fouling was caused by deposition of corrosion byproducts (Table 12.8). In addition to the maintenance corrosion control treatment, a polyacrylate dispersant to disperse the iron was dosed at 200 mL/L. The biocide dosing was increased to daily slug additions of 50 mL/L, based on the volume of water in the system.

Three months later another inspection revealed tubercles as well as extensive masses of corrosion byproducts blocking the tubes in the heat exchanger. The tubercles, nodules, and stalks were typical of those produced by the uncontrolled growth of *Gallionella sp.* (iron-oxidizing bacteria) and *Beggiatoa sp.* (iron- and sulfur-oxidizing bacteria). The microbiological population had remained high, and the pH of the circulating water had dropped to 7.4, with further sporadic downward excursions to 6.0 and 6.5 for no apparent reason.

As the online cleaning treatment program was initiated, the pH was lowered to and maintained at 4.5 for five days. During that period, the phosphonate addition was increased substantially, and the microbiocide dosage was increased to 100 mg/L daily. This proved unsuccessful. Even though the online cleaning had reduced the microbiological total plate count to zero temporarily, one week later the total plate count had increased to more than 10 million colonies per mL of aerobic bacteria. Corrosion byproducts continued to deposit in the system and contribute to fouling (Table 12.8). Throughout this period, the weekly microbiological survey did

TABLE 12.8

Chemical Analysis of Corrosion Byproducts

Byproduct	Percent of Ash
Free Iron as Fe	Nil
Iron as Fe_2O_3	89.40
Chromium as Cr_2O_3	0.05
Zinc as ZnO	0.04
Nickel as Ni_2O_3	0.01
Aluminum as Al_2O_3	0.04
Copper as CuO	0.10
	Percent of OD deposit
Oils and greases	5.00

not detect anaerobic SRB.

The microbiocide dosage was maintained at 100 mL/L, applied daily. Another dispersant treatment was tried, replacing the phosphonate. The replacement product was a formulation of two polyacrylates combined with an aminotris (methylene-phosphonic acid). The dispersant was dosed at a concentration of 500 mg/L based on the volume of makeup to the system. Thereafter, weekly inspections were made concurrent with the microbiological survey. After three weeks, it became apparent that this particular dispersant treatment was preventing further deposition of corrosion byproducts (Figure 12.24, p. 200). However, accelerated corrosion rates, high microflora populations, and unpredictable pH excursions persisted.

Definition of Problem

To mitigate problems, a procedure was established to define the problems and determine causes so corrective action could be taken.

Investigative Steps

Review of Past Observations

It was obvious the occurrence of increased corrosion and fouling was related to the change in makeup water source and to the contamination of the system with oil. The original makeup and circulating water chemical properties were characterized as being corrosive (i.e., Ryznar Stability index greater than 9.5).

However, the corrosion rates were acceptable (i.e., less than 25μm/y), and fouling by corrosion byproducts was not significant. This situation was attributed to the effectiveness of the zinc-chromate corrosion inhibitor application and the phosphonate dispersant treatment. The original relatively low microflora populations were attributed to operating conditions; the closed-loop system did not support unrestricted growth of problem-causing microorganisms. It was suggested the periodic addition of the organosulfur microbiocide may have restricted the occurrence of SRB.

No significant changes occurred in the gross chemical characteristics of makeup or circulating water when recovered waste water was commingled with well water.

Compatibility of Treatment Chemicals

Since one of the changes at the onset was increased microbiocidal dosage, the next step in the investigation was to verify the compatibility of the organosulfur microbiocide with the chromate-based corrosion inhibitor. The system's circulating water, containing the corrosion inhibitor treatment, was used as a test medium in laboratory studies in which use-concentration dosages of the microbiocide were added to the water. Analysis of the test medium treated with microbiocides indicated the chromate residual maintained in the circulating water had been reduced from the hexavalent to the trivalent form soon after the addition of the microbiocide. It was then determined the trivalent chromate did not inhibit corrosion adequately under the corrosive operating conditions. Another factor limiting the effectiveness of the corrosion inhibitor was sludge formed by interaction of the chromate-based inhibitor and oil that had contaminated the system.

Assessment of Dispersant Chemicals

Visual inspections and photographs were made of the circulation system and heat exchange surfaces. Subjective comparisons were made between the performance of the phosphonate and the phosphonate/acrylate dispersants based on the following parameters:

- Deposits on surfaces of system piping;
- Formation of tubercles, nodules and stalks;
- Deposits on heat transfer surfaces;
- Sludge accumulation in low-velocity flow sites; and
- Plugging of tube inlets at the heat exchangers.

As stated earlier, the performance of the phosphonate dispersant adequately inhibited scale formation and fouling when well water was available for makeup. When waste water was used as part of the makeup, the phosphonate at increased concentrations did not adequately prevent deposition or fouling.

From the onset of the use of the phosphonate/acrylate dispersant, a decrease was observed in the amount of surface deposition of the system piping and heat exchange surfaces. Sludge accumulation and plugging of heat exchanger tubes were minimized.

Characterization of the Microflora

The coincidence of high microflora populations and the occurrence of the corrosion and plugging problems validated the need to determine the types of microorganisms constituting the total microflora. Characterization was achieved with direct microscopic examining techniques and traditional microbiological culturing procedures. These were supplemented by extensive inspections of the system. Special note was made of changed properties of the circulating water. These investigations revealed the following observations:

- Extensive formation of tubercles, nodules and stalks,
- Presence of a black fluid under and within the tubercles,
- Presence of sulfur granules within cells of some bacteria found in the deposits,
- Inexplicable drops in pH,
- H_2S evolution on acid addition to deposits,
- FeS found on heat exchanger surfaces under deposits,
- Pitting of metal surfaces under tubercles and sludge deposits,
- Presence of ferric oxide stalks and "rosettes" typical of those associated with iron-oxidizing bacteria of the genera *Gallionella* and *Caulobacter*.

Through selective media culturing techniques, the following genera were positively identified as being part of the microflora in the circulating water and sludge. Some of the specific characteristics of each genera observed in the system are included here:

1. *Gallionella sp.*—Referred to as the stalked iron bacterium; aerobic; curved rod that forms hydrous ferric oxide strands in twisted or straight bands; observed microscopically (Figure 12.25).

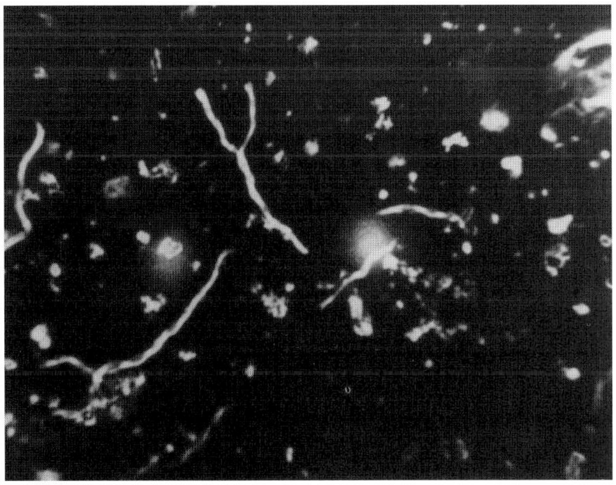

FIGURE 12.25
Microscopic observation of typical strands of hydrous ferric oxide associated with growth of *Gallionella sp.* (900X original mag.)

2. *Sphaerotilus sp.*—Iron-oxidizing rod bacterium that typically forms a sheathlike filament, encrusted with ferric oxide or hydroxide. The organism was observed by microscopic examination to develop an extremely dense biomass that entrained dirt and debris, forming plugging deposits and creating anaerobic environment under the deposits (Figure 12.26).

3. *Thiobacillus sp.*—Acidophilic aerobic rod bacterium that oxidizes elemental sulfur and

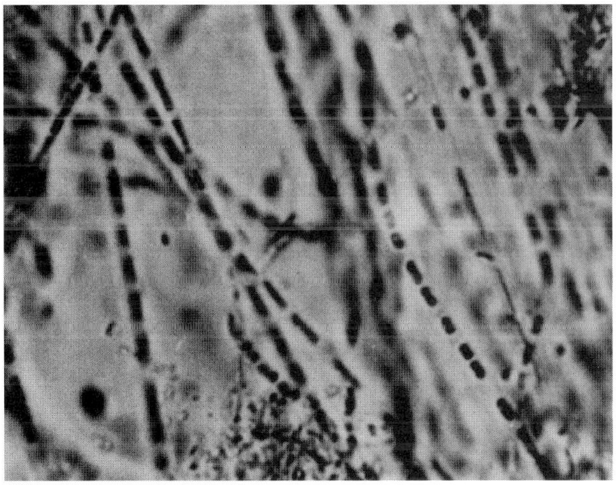

FIGURE 12.26
Microscopic observation of typical filaments of *Sphaerotilus sp.* (1000X original mag.)

metal sulfides to produce H_2SO_4 as a metabolic byproduct. Low pH conditions were associated with the growth of this organism.

4. *Desulfovibrio sp.*—Obligate anaerobic motile spirillum bacterium that obtains its energy by reducing sulfates to sulfides, producing H_2S and metal sulfides as metabolic byproducts. This mechanism typically produced pitting corrosion in the anaerobic environments under tubercles and other deposits found on metal surfaces in the system (Figure 12.27).

Based on observations during this study and interpretations of information obtained from the investigation described above, the problem was defined as MIC, manifested by extensive pitting and high corrosion rates. The plugging and fouling, a major part of the problem, were a result of the deposition of the MIC byproducts. Interpretation of the information indicated the problem was caused by the ingress of oil into the system and the subsequent use of makeup water that supported several types of troublesome microorganisms. The absence of an effective microbiocidal treatment procedure promoted the uncontrolled growth of the microflora. This situation subsequently allowed specific genera of bacteria, capable of inducing corrosion, to become dominant members of the microflora.

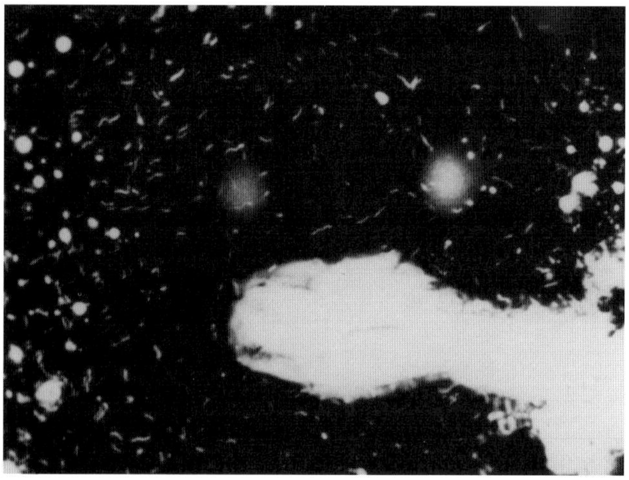

FIGURE 12.27
Microscopic observation of SRB, *Desulfovibrio sp.* (1500X original mag.)

Once the corrosion mechanism was established, use of a microbiocide treatment alone was not adequate to mitigate the corrosion. The use of chromate-based corrosion inhibitor was ineffective because of its incompatibility with the microbiocide. Moreover, for a period of time, oil contamination interfered with this inhibitor. The phosphonate could not keep depositable material dispersed under existing operating conditions.

Mitigation of the Problem

By defining the problem and cause, it was possible to develop a procedure to overcome it. A different corrosion inhibitor treatment was necessary, and the use of an effective dispersant was critical to achieve three objectives necessary to mitigate the problem:

- Prevent the deposition of corrosion byproducts,
- Keep the solids dispersed to prevent the creation of conditions ideal for microbiological growth, and
- Clean surfaces at established corrosion sites so the corrosion inhibitor would be capable of passivating the system.

Understanding the definition and causes of the problem reinforced the idea that the individual segments of the chemical treatment program must be compatible. If possible, each component should complement the effectiveness of the others. Additionally, two other considerations were necessary:

- Eliminate chromates because of their adverse environmental impact, and
- Develop a treatment program that would be effective even with makeup water of extremely low quality.

Both numerous visual inspections and the corrosion rate data obtained by monitoring the system substantiated an extensive amount of corrosion within the system. The established corrosion sites must be passivated by removing the corrosion byproducts from the system as much as possible and by using a corrosion inhibitor (passivator) that is effective on less than totally clean metal surfaces.

The online cleaning procedure attempted had

been unsuccessful. It appeared that low pH did not contribute to the removal of the corrosion byproducts or fouling materials. Therefore, no progress was achieved by cleaning the system.

Selection and Application of Online Cleaning Treatment

Successful experience using the phosphonate/acrylate dispersant was the determining factor for its use as part of the cleanup procedure. A review of the corrosion rate data confirmed that its use was not associated with an increased corrosion rate.

The presence of residual oil contamination and the substantial amounts of biomass in the system determined the use of penetrant/biodispersant with the dispersant. After reviewing the characteristics of the available compounds and the required specifications, a compound composed of dimethylamide of a fatty acid and sodium-2-mercaptobenzothiazole was selected. This provided both the penetrant/biodispersant requirements and a barrier film corrosion inhibiting mechanism. Features included:

- Function over a wide pH range, not pH-dependent;
- Total compatibility with most microbiocides (enhances effectiveness of nonoxidizing biocides);
- Biodegradable effluent, no adverse environmental impact;
- Effectiveness as film-forming corrosion inhibitor in presence of oil contaminants (keeps oil dispersed in water);
- Does not contribute to thick films or deposits (forms monomolecular film on metal surfaces); and
- Penetration through corrosion byproducts on metal surfaces and physically passivates active corrosion sites.

The selected compounds were added to the closed-loop circulating water after flushing the system and refilling with makeup water containing presoftened waste water. Monitoring procedures described earlier were done, and periodic visual inspections made. Photographs were taken of selected sites in the system that had previously been problem areas (Figure 12.24, p. 200). Initial dosages of the compounds, based on the volume of water in the system, were 200 mg/L of phosphate/acrylate dispersant and 100 mg/L of dimethylamide/2-MBT corrosion inhibitor. These concentrations were maintained by adding more chemicals at the required dosages based on volume of makeup added to the system after initial makeup.

From just prior to the initiation of the treatment to two months after its start, the corrosion rates dropped from more than 75 µm/y to less than 20 µm/y measured by linear polarization electrodes and to 25 µm/y measured by coupon weight loss. From the start of the treatment period, the heat flux and operating efficiencies of the system were monitored routinely. At no time were there losses in heat transfer efficiencies, nor did the heat flux deviate from the design specifications of the system.

Selection and Application of Microbiocide Treatment

Routine total microflora plate counts and SRB counts were continued weekly following the initiation of the cleanup chemical treatment. During this treatment period, the organosulfur microbiocide was added to the system three times a week at a slug dosage of 100 mg/L, based on volume of water in the system. The results in Table 12.6 show the total counts for aerobic bacteria continued to exceed 1×10^6 colonies per mL and the SRB counts continued to be high. The unpredictable pH variations continued as well.

Based on this, work was initiated to determine if a more effective microbiological treatment program could be developed. Samples of the biomass were collected from the system for use as inoculum in a biocide selection screening test. Actual circulation water from the system was used as the test substrate.

Results of the selection test indicated the most effective treatment program was two nonoxidizing microbiocides. Chlorination was not considered because of the high pH of the circu-

TABLE 12.9

Microbiological Survey Data for Period Immediately Before and After Dosing Circulating Water with Microbiocides

Time	Aerobic Bacteria	Total Counts- Colonies per mL anaerobic SRB	Molds and yeasts
June 19, 1984 @ 14h38 (pre-dosing)	2,420,000	Positive 22 colonies counted	Nil
June 19, 1984 @ 19h47	540,000	Nil	Nil
June 20, 1984 @ 09h00	Nil	Nil	Nil
June 20, 1984 @17h15	Nil	Nil	Nil
June 21, 1984 @12h00	Nil	Nil	Nil

lating water and its own potential contribution to increased corrosion rates. Microbiocides selected were compound (a) sodium-dimethyldithiocarbamate, 40 percent active; and compound (b) methylene bis(thiocyanate), 10 percent active, plus 2-(thiocyanomethylthio)benzothiazole, 10 percent active.

These were used simultaneously, not alternately. Dosage rates based on the volume of water in the system were 50 mg/L of compound (a) and 20 mL/L of compound (b).

The results from the initial slug treatment are shown in Table 12.9. Two days after the treatment the microbiological survey results indicated no detectable levels of microflora in the system. Weekly dosage of these compounds has continued to maintain the microflora population at acceptable levels. The decrease in the microbiological population has coincided with the ceasing of pH deviations of the system's circulating water. The coordinated treatment program led to a significant decrease in the corrosion rate. Physical inspection of the system revealed a marked improvement in the surfaces of the heat exchanger and a significant decrease in the volume of sludge generated.

Summary

Based on the time taken for this work, the use of the dimethylamide in conjunction with a polyacrylate/phosphonate dispersant to control electrochemical corrosion and to maintain metal surfaces deposit-free seems to be extremely effective. To effectively control total corrosion in the system, it is believed that an efficient corrosion inhibition program must be used in conjunction with effective microbiological control. Without this, the program could be useless.

This was the approach taken in treating the bloomcaster closed systems. The results have been effective; however, the treatment is more expensive than the chromate-based treatment. There is currently no reliable method for monitoring the concentration of dimethylamide in the circulating water. This could lead to a control problem since the product must be used volumetrically.

The positive points of the treatment include clean systems, low corrosion rates, and oil dispersion by the product if contamination occurs. The stepwise procedure outlined documents that a severe corrosion problem induced by microorganisms can be mitigated and controlled.

Because of the success attained here, the same approach has been used on the bloomcaster open system that experienced similar problems. This treatment has proven equally successful.

Acknowledgments

Thanks go to Mrs. P.J. Allison, chief microbiologist of Buckman Laboratories (Pty) Ltd., Hammarsdale, South Africa, and to the staff in the

South African Iron and Steel Corp.'s reprographic department for assistance with illustrations.

References

1. R.M. Atlas, R. Bartha, *Microbial Ecology Fundamentals and Applications* (Reading, MD: Addison-Wesley Pub. Co., 1981).
2. R.W. Lutey, CORROSION/80, paper no. 39 (Houston, TX: NACE 1980).
3. M.J. Pelczar, Jr., R.D. Reid, *Microbiology*, 2nd ed. (New York, NY: McGraw-Hill Co., 1980).
4. J.M. Sharpley, "Microbiological Corrosion and Its Control," *Corrosion Inhibitors* (Houston, TX: NACE, 1973).
5. R.Y. Stainier, E.A. Adelberg, J.L. Ingraham, *Microbial World*, 4th ed. (Englewood Cliffs, NJ: Prentice-Hall, 1976).

Microbiological Corrosion Problems in a Paper Mill Closed Water System[4]

By A. Piluso

This case describes corrosion that occurred from a recovery process involving "white water" in a paper mill. White water is the relatively clean water that carries the pulp to the paper machine. This water passes through the Fourdrinier screen as the pulp lays down on it and traditionally is discharged to the nearest river. Sulfur compounds from the process, together with hydrogen sulfide and organic acids produced by bacteria, give this outfall a foul odor.

Zero discharge is being attained by many progressive mills by collecting white water, passing it through filters and screens to catch much of the suspended fibers, then cleaning it further in a clarifier. Clarified white water and solids are recycled back to the process. Economic benefits are realized in higher fiber yield and reduced treatment costs for fresh makeup water.

At one such mill, a multicylinder machine produces 170 tons per day of specialty paper, including box board, chip board, and candy boxes. White water is collected in a 7,600 m^3 (two million gal.) clarifier where retention time averages approximately 20 h. The pH is 6 to 7, and temperatures are 40 to 46°C (105 to 115°F), conditions ideal for growth of bacteria and other microorganisms.

Types of corrosion seen in the system are shown in Figures 12.28-12.31. Figure 12.28 (p. 201) shows general corrosion of a mild steel wire in the clarifier. This occurred under a heavy buildup of fiber sludge which typically contains a high level of inorganic residues as well as numerous microorganisms.

Figures 12.29 and 12.30 (p. 201) show corrosion of a large AISI 304 stainless steel pump impeller. Figure 12.31 (p. 202) shows

[4]Originally published in *Materials Performance* 20, 8(1981): p. 42

FIGURE 12.29
AISI 304 SS pump impeller showing irregular corrosion that occurred under bacteria deposits.

severe bacteria induced corrosion of a AISI 304 stainless steel screen located ahead of the paper machine.

In addition to the corrosion problems, noxious gases, including hydrogen sulfide, methane and mercaptan, were generated in the clarifier. These were carried through to the finished product, causing customer complaints and product rejections. Further complications resulted from the generation of organic acids, including acetic and butyric. These affected the solids in the clarifier, causing them to stay in suspension or float to the top rather than settling to the bottom, where they could be removed and recycled.

Corroded areas were typically covered with aerobic slime deposits containing *Pseudomonas*, *Aerobacter*, *Flavobacterium*, and/or *Bacillus* bacteria. Sometimes these were accompanied by a process fiber buildup. Under the slime layer, there was generally a brown, gritty deposit that might contain one or more of the anaerobic bacteria *Desulfovibrio*, *Desulfotomaculum*, and *Clostridium*. The latter is known to ferment cellulose to acetic acid and ethanol, and also can convert sugars to butyric acid.[1] Both *Desulfotomacula* and *Clostridia* are spore formers, with the spores able to survive the extreme temperatures of the paper dryer and thus carry over into the finished product. In addition to discrete deposits, the clarifier bottom sludge was rich in anaerobic bacteria, which are recycled to the process.

Control of this problem was difficult because any chemical additions must be compatible with the process and not affect the final product. A thiocyanate-type biocide reduced odors and customer complaints, but did little to control overall bacteria counts in the system. Returning the clarifier sludge more continuously to the process rather than allowing it to accumulate is expected to reduce bacteria growth somewhat. Forced aeration of the clarifier has been proposed to further inhibit growth of the anaerobes. Better housekeeping to minimize process deposits will help a great deal.

It seems inevitable, however, that corrosive bacteria will always be part of the closed papermaking process, and their effects will have to be minimized by measures outlined and, in some cases, use of more exotic materials of construction.

Reference

1. J.W. McCoy, *Microbiology of Cooling Water* (Chemical Publishing Co., 1980).

Corrosion in a Waste Water Treatment Tank[5]

By G. Kobrin and R.L. Strong

Waste water treatment at one DuPont plant in the Gulf Coast area is a three-step process:

1. Microbiological oxidation of organic carbon forms to carbon dioxide, and organic nitrogen forms to ammonia,
2. Microbiological nitrification, with aeration, of ammonia forms to nitrates, and
3. Clarification to remove microorganisms for return to Steps 1 and 2.

This case history involves severe corrosion of the Step 2 nitrification tank after only six months of service and includes a discussion of probable corrosion mechanisms, repairs and steps taken to prevent recurrence.

Materials of construction for the 43 m (140 ft) diameter open-top tank were:

- Bottom — reinforced concrete,
- Sidewall and internal waste water piping — carbon steel coated with an inorganic zinc primer/catalyzed epoxy topcoat system,
- Internal air distribution piping and supports — AISI 304 (SS), and
- Internal catwalk and structural supports — galvanized steel.

After only six months of service, the tank started leaking through a carbon steel channel that joined the sidewall to the bottom. The tank was drained and cleaned for entry and inspection. A layer of sludge approximately one foot deep was found on the bottom after draining, but before cleaning.

The tank interior sidewall-to-bottom junction is shown in Figure 12.32. On initial inspection, the channel presented a fairly uniform rusted appearance with no indication of pitting or perforation. However, the rusted areas actually were dense, nonmetallic deposits that could be carved out with a pocket knife, exposing extensive, deep local attack that penetrated the full channel thickness of about 6 mm (1/4 in.) at several locations. One such location is shown in Figure 12.33 (p. 202).

In the authors' experience, this attack is most unusual and not at all typical of aqueous corrosion of carbon steel. The *in situ* wastage of the channel resembles the well-known graphitic corrosion (graphitization) of cast iron, where the iron matrix is preferentially attacked, leaving a soft plug of graphite. Followup metallographic examination of a channel section showed nothing unusual in the microstructure that would account for this attack. Unfortunately, the deposits from the severely corroded and perforated areas were discarded before samples could be obtained for analysis.

Figure 12.34 (p. 203) is typical of the appearance of the AISI 304 SS air distribution piping network, which was positioned a few inches above the concrete bottom and found to be partially immersed in the aforementioned sludge layer. There were voluminous moundlike deposits, particularly along the bottom half. These deposits were heavier and more concentrated in the relatively stagnant locations between the black vertical air/water mixers also shown in this figure.

---

[5]Originally published in *Materials Performance* 20, 8(1981): p. 46.

A portion of an air distribution pipe, after removal of deposits by sandblasting, is shown in Figure 12.35 (p. 203). The broad deep pits were located beneath the deposits shown in Figure 12.34 (p. 203). Many of these pits contained concentric rings, typical of corrosion of carbon steel under SRB colonies as reported extensively in the literature. Pitting had perforated the stainless steel piping at numerous locations.

The AISI 304 stainless steel air distribution piping supports (Figure 12.36, p. 204) were located in relatively turbulent areas beneath the air/water mixers and showed little or no attack. However, several support anchor bolts embedded in the concrete bottom were almost completely consumed by corrosion. The corroded bolts were found to be sulfur-bearing AISI 303 stainless steel; noncorroded bolts were AISI 304 as originally specified.

Internal carbon steel waste water piping showed severe localized corrosion and perforation at numerous locations. An example of this attack is shown in Figure 12.37 (p. 204). No corrosion occurred where the coating was intact. This condition, but to a lesser extent, was also found on the carbon steel tank sidewall. It is interesting that once the epoxy topcoat was perforated, the underlying inorganic zinc primer provided no galvanic protection to the underlying steel. This condition was even more apparent on the internal galvanized steel structural members. A severely corroded galvanized support for piping is shown in Figure 12.38 (p. 205). It is possible zinc is cathodic to carbon steel in this environment. Zinc seems to pay a similar role in Case History CS-1.

The only tank component essentially free from deterioration was the reinforced concrete bottom.

We have attributed corrosion of the various components to several mechanisms, including:

- Microbiological organisms, both anaerobic and aerobic;
- Galvanic couples;
- Oxygen concentration cells; and
- Acid concentration cells.

Microbiological analysis of a sludge sample and wet corrosion products from several locations showed high counts of SRB and *Sphaerotilus* iron bacteria along with a multitude of other microorganisms. Corrosion products from the SS air distribution piping and carbon steel piping were rich in sulfides. Sludge samples showed pH between 3 and 4, whereas the waste water normally was close to pH 7.

These repairs were made:

1. The carbon steel channel which supports the tank sidewall at the reinforced concrete bottom was sandblasted, corroded areas were repaired by welding, and the entire sidewall-to-bottom junction was covered with a heavily reinforced epoxy resin thick film lining system.

2. Corroded areas in the carbon steel piping and sidewall were sandblasted, weld repaired where required, and spot coated with a proprietary glass flake reinforced epoxy thin film lining system.

3. The entire stainless steel air distribution piping network was removed from the tank, sandblasted, weld repaired as required, and completely coated with the aforementioned thin film lining system.

4. All galvanized steel structural support members were sandblasted, weld repaired as required, primed and completely encased in reinforced concrete.

5. An impressed current cathodic protection system, complete with rectifier and high silicon iron anodes, was installed and adjusted to -0.95 volt minimum structure-to-waste water potential (vs $Cu/CuSO_4$).

After eight months of additional service, the tank was drained and cleaned for inspection. All internal components were in excellent condition. A few blisters were found in portions of the original sidewall coating. The blisters were filled with water at pH 13; underlying steel showed little or no attack. These areas were spot sandblasted and recoated with the aforementioned thin-film lining system.

Waterside Corrosion of Monel Heat Exchanger Tubes[6]

By J.G. Stoecker

A nickel-copper alloy 400 (UNS NO4400) heat exchanger was removed from process service because of a decline in heat transfer efficiency. A routine shop hydrotest of the unit before cleaning revealed many leaks in the 15.9 mm (0.625 in.) OD x 12 mm (0.049 in.) wall tubes. This finding prompted a complete examination of the unit.

The examination revealed that several tubes were fouled and partially plugged with calcium phosphate scale on the tube or treated water side of the exchanger. All tubes had evenly distributed localized corrosion cells on the inside surface, as shown in Figure 12.39 (p. 205). Each cell was covered with a vivid green-colored deposit. The deposits ranged in size from 0.8 to 4.8 mm (0.03 to 0.19 in.) in diameter. The average penetration of the cells reached a depth equivalent to approximately 75% of the tube wall thickness. Metallographic examination found extensive intergranular corrosion in each cell (Figures 12.40 and 12.41); in addition, selective leaching of nickel, leaving a spongy, copper-rich material beneath the deposit, was evident in some cells.

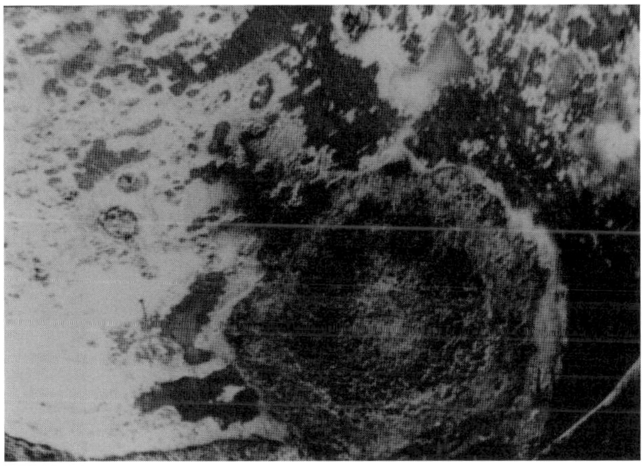

FIGURE 12.40
SEM micrograph of corrosion cell on inside surface of Monel tube.

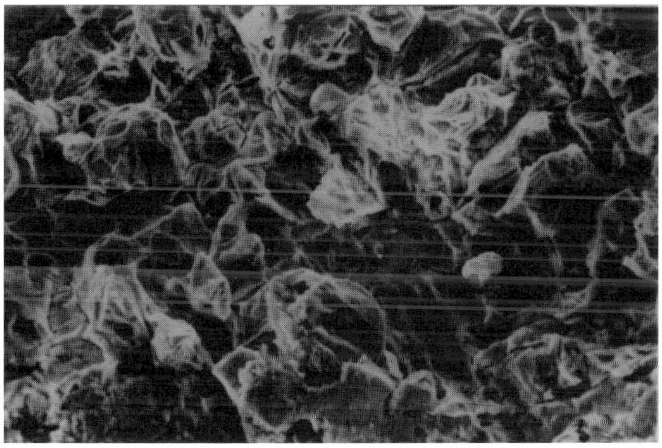

FIGURE 12.41
SEM micrograph of corroded surface inside cell in Monel tube.

Auger microprobe analysis of the alloy 400 tube (Table 12.10) revealed an alloy composition by weight of 0.1 to 0.3% carbon, 27.4 to 28.3% copper, and 71.6 to 72.3% nickel. Tables 12.11, 12.12, and 12.13 show the composition of the deposits on the inside surface of the tubes by x ray diffraction, chemical analysis of water solubles, and spectrographic analysis procedures.

Results of the chemical analysis tend to support the observation of the selective leaching of nickel because there is a higher nickel content in the water-soluble portion of the corrosion deposits; however, copper was also attacked. There was evidence on some deposits

[6] Originally published as CORROSION/83 paper no. 245 (Houston, TX: NACE, 1983).

TABLE 12.10
Auger Microprobe Analysis of an Alloy 400 Tube

Analysis Location	Carbon (wt%)	Copper (wt%)	Nickel (wt%)
A (base metal)	0.3	27.6	72.1
B (base metal)	0.1	28.3	71.6
C	0.3	27.8	71.9
D	0.3	27.4	72.3
E	0.3	27.6	72.1

TABLE 12.11
X-ray Diffraction Analysis of Deposits on Inside Surface of Alloy 400 Tubes

Cu_2O	Major
$Ca_5(PO_4)_3(OH)$	Major
$CaSO_4$	Minor
Major:	>10%
Minor:	1 to 10%

TABLE 12.12
Chemical Analysis of Water-Soluble Portion of Deposits on Inside Surface of Alloy 400 Tubes (μg/g)

Cl-	450
PO_4^{-3}	725
SO_4^{-2}	2,450
Cu	404
Ni	1,224

that copper removed from the metal matrix was reduced to elemental copper, which adhered to the underside of some deposits.

The results of these analyses are of particular interest from a microbiological standpoint. The sulfate level in the deposits in a system that is supplied by a relatively low sulfur and sulfate water source suggests sulfur or sulfate microbiological activity. Increasing the sulfur content in a system generally increases microbial growth, and

TABLE 12.13
Spectrographic Analysis (Semi-Quantitative) of Deposits on Inside Surface of Alloy 400 Tubes

Element	wt%
Arsenic	0.02
Bromide	0.05
Calcium	6.00
Chloride	0.50
Cobalt	0.02
Copper	42.00
Iron	3.00
Magnesium	0.20
Manganese	0.10
Nickel	16.00
Phosphorus	3.00
Silicon	0.50
Sodium	0.20
Strontium	0.10
Sulfur	0.30
Zinc	0.30

NOTE: pH = 7

grow th is retarded when sulfur concentrations are less than 0.01%, as reported by Bennett.[1] Hard water, 692 ppm $CaCO_3$, has been shown to produce fast growth of *Pseudomonas* species. For those interested, the literature contains much information concerning the effects of various inorganic salts on the growth of many microbe species.

It is difficult to believe that treated freshwater at pH 7 and moderate temperature would pit alloy 400 to the extent observed in a matter of months. The selective leaching of nickel from this alloy by a completely aqueous system would also seem improbable without some outside influence. MIC is definitely indicated in this case history, but can never be totally proven with the background data available.

Reference

1. E. Bennett, *Lubrication Engineering* July (1972).

Involvement of a Thermophilic Bacterium in Corrosion Processes[(7)]

By B. Little, P. Wagner, S.M. Gerchakov, M. Walch, and R. Mitchell

Abstract

A rapidly proliferating obligate thermophilic bacterium with an optimum temperature range of 60 to 70°C (140°F to 158°F) was isolated from a failed nickel 201 brazed joint. Corrosion experiments have demonstrated that the isolate is responsible for the *in situ* production of acidic metabolites and creation of differential aeration cells, resulting in a significant corrosion current.

Introduction

MIC has been recognized by corrosion engineers and scientists for more than 50 years. Review articles and books by Miller and Tiller,[1] Iverson,[2] McCoy,[3] Tatnall,[4] and Pope[5] document the problem. Despite this recognition and documentation, the identification of specific mechanisms for MIC has remained elusive because of the complexity of microbiological processes and lack of analytical techniques.

Polarization techniques, traditionally used for measuring corrosion rates in aqueous environments, assume that a metal is corroding by a Wagner-Traud[6] mechanism, i.e., that anodic and cathodic processes occur with equal probability on all parts of the corroding surface. The metal surfaces colonized by microorganisms are not homogeneous. Instead, anodic and cathodic areas are distinct in space and stable in time, which results in localized corrosion.

In this paper, a corrosion-measuring apparatus is described that permits separation of the anode and cathode and evaluation of the electronic processes. The apparatus has been used to systematically investigate the mechanisms by which thermophilic bacteria can be shown to accelerate corrosion.

Experimental Procedures

Experiments were designed to accelerate the corrosion failure of braze joints by elevating temperature. Replicate 1.4 in. (0.635 cm) inside diameter nickel 201 tubing joints brazed with alloy BNi-3 (UNS N99630) were maintained at room temperature, 60° and 80°C (140° and 176°F) with distilled water at a flow rate of 1 ms^{-1}. The wall thickness of the tubing and the brazed area was 0.040 in. (0.1 cm). It was predicted that at elevated temperatures, the rate-determining step in the corrosion process would be accelerated and that failures would result first in the systems maintained at 80°C, followed by those maintained at 60°C, and finally those at 20°C. After 21 days of continuous operation, the first leak was observed in the braze fillet maintained at 60°C.

Isolation of Bacteria

Bacteria were isolated from the failed nickel joint by incubating pieces at 65°C in a basal salts medium supplemented with 0.1% tryptone and 0.1% yeast extract. The salts medium contained (per liter) 100 µg nitrilotriacetic acid; 2.8 µg $FeCl_3$; 60µg $CaSO_4 \cdot 2H_2O$; 22 µg $\cdot$ $MnSO_4 \cdot H_2O$; 100 µg $MgSO_4 \cdot 7H_2O$; 5 µg $ZnSO_4 \cdot 7H_2O$; 8 µg

---

[(7)]Originally published in CORROSION 42, 9(1986): pp. 533-536.

NaCl; 5 µg H_3BO_3; 103 µg KNO_3; 0.16 µg $CuSO_4$; 689 µg $NaNO_3$; 0.25 µg $Na_2MoO_4 \cdot 2H_2O$; 111 µg Na_2HPO_4; and 0.46 µg $CoCl_2 \cdot 6H_2O$. The final pH value of the supplemented medium was adjusted to 7.6 with NaOH, and the medium was sterilized by autoclaving at 121°C, 15 psi (100 kPa), for 15 min.

After two days of incubation without shaking, the broth became turbid, and transfers were made to fresh broth. Subsamples of the turbid broth were streaked onto a solid medium of the same composition plus 3% agar. Isolated colonies were picked and restreaked to obtain a pure culture.

Organisms attached to metal surfaces were selected by providing a solid substratum (a sterile nickel disc) in the culture tube with the culture medium. The first such tube was inoculated with a pure culture of the isolate. Subsequent inoculations were made by transferring the nickel disc to a fresh tube containing a sterile medium and another sterile nickel disc. In this way, only organisms attached to the nickel were transferred. Maintenance of the cultures and subsequent corrosion experiments used the above basal medium with 0.025% tryptone and yeast extract.

Scanning Electron Microscopy

Brazed sections were removed from the accelerated corrosion test setup, and the flow-through water was replaced with cacodylate buffered 4% glutaraldehyde, which remained in the section for 4 h. This solution was followed by successive washes with distilled water, acetone, acetone and xylene, and xylene. The air-dried specimens were held in a vise and sectioned with a standard high-speed metal cutting saw (13 teeth cm^{-1}) before being sputter coated for viewing with an Amray† (Model 1000 A) scanning electron microscope. Nickel electrodes from the corrosion measuring experiments were prepared for scanning electron microscopy (SEM) in a similar manner.

Measurement of Organic Acids

To test for organic acid production by the isolated bacterium, cells were grown for 24 h at 65°C in the medium described above, containing 0.1% each of tryptone and yeast extract. Samples of this culture (0.5 mL) were analyzed for volatile and nonvolatile fatty acids by gas chromatography.[7] For volatile acids, the samples were acidified and extracted with ether. For nonvolatile acids, methyl derivatives were prepared and extracted in chloroform. A stainless steel column (1.83 m x 0.318 cm) containing 10% SP-1000 on 100/120-mesh Chromosorb† was used on a Hewlett-Packard 5700A† gas chromatograph equipped with a flame ionization detector. The acids were identified and quantified by comparing the retention times and peak heights with those of known standards.

Corrosion Experiments

A schematic diagram of the corrosion measuring device is presented in Figure 12.42. Ground glass bases from 47-mm Millipore† filter supports were joined to the sides of two 500-mL flasks. A 0.1-µm pore Millipore cellulose acetate/cellulose

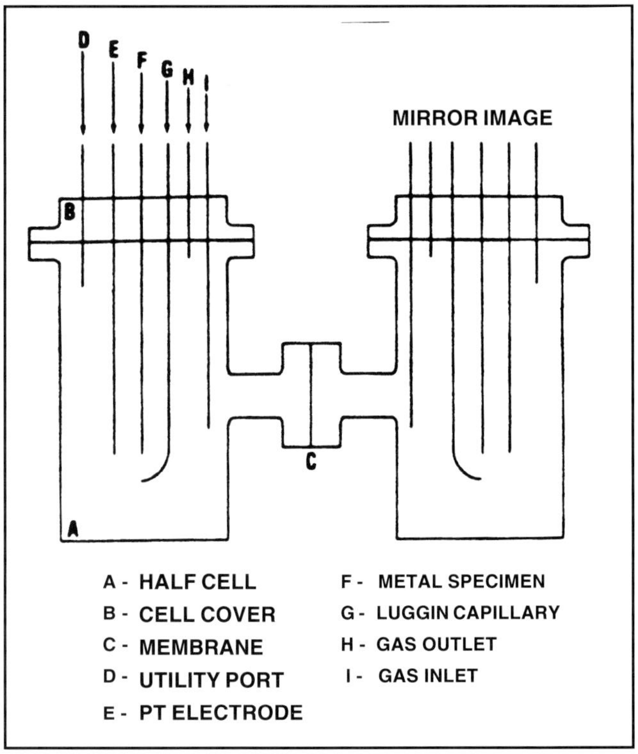

FIGURE 12.42
Diagram of the corrosion measuring device.

†Trade name.

nitrate filter disc was placed between the two supports, and the flasks were clamped together and filled with 500 mL of the basal salts medium supplemented with 0.025% each of tryptone and yeast extract. An EG&G flat electrode holder containing a 1 cm^2 flat disc of nickel 201 was placed into each flask. The entire assemblage was autoclaved for 15 min at 120°C (248°F) and 15 psi (100 kPa). After cooling, the flasks were placed in a water bath aquarium maintained at 60°C (140°F). Water-saturated air was bubbled into one of the flasks.

Previous experiments have revealed that oxygen diffused across the membrane and the water in the two flasks was saturated with oxygen after 2 h of aeration. The membrane ensured electrolytic conductivity between the discs while maintaining microorganisms on one side of the membrane and an abiotic system on the other side. One flask was inoculated with a nickel disc previously colonized by microorganisms while a sterile disc was dropped into the flask to be kept in an abiotic condition. The electrodes were then coupled, and the resulting currents were measured with an EG&G PARC† (Model 350) corrosion measuring device.

In such a system, the measuring device functions as a zero resistance ammeter; i.e., the potential of the cell is not perturbed by the current measurement. If two identical metal specimens are coupled and isolated in identical electrolyte solutions, the current flow between them is zero, even in the presence of active corrosion. They corrode at the same rate, and neither functions as an anode or cathode in reference to the other. If this balance is disturbed by the presence of colonizing microorganisms, a current flows between the two specimens. If the reaction occurring at the surface of the working electrode is oxidation, an anodic current will be recorded; if it is reduction, a cathodic current results.

Experiments were designed to test the individual effects of bacterial respiration and metabolites on the corrosion of nickel 201. The 0.1-µm Millipore filter of the above system was replaced with a TFE-fluorocarbon disc to ensure chemical separation of the compartments. A KJC1/agar bridge maintained the electrolytic conductivity. The effect of oxygen concentration was tested by bubbling nitrogen in one flask while bubbling oxygen in the other. To evaluate the effects of acid metabolites, various concentrations of isobutyric and isovaleric acids were added to the medium of one of the flasks. These experiments were conducted at 23° and 60°C (73° and 140°F).

†Trade name.

Results and Discussion

Increasing the temperature to affect an increase in corrosion rate assumes that the rate-determining step in the overall corrosion process is the activation process. In general, if diffusion rates are doubled for a certain increase in temperature, activation processes can be expected to increase 10 to 100 times, depending on the magnitude of the activation energy.[8] In deaerated solutions, when the rate-determining step is under activation control, the primary effect of increasing the temperature is to increase the exchange current.

Conway et al.[9] found that for nickel the exchange current increased from ~ 10^{-2} to 1.0 A/m^2 when the temperature increased from 10° to 75°C (50° to 167°F), and that the activation energy was ~ 59 kJ. Temperature increase is complex in that the diffusivity of oxygen molecules increases, but solubility decreases. The net transport of oxygen should increase with temperature until a maximum is reached (estimated at 80°C [176°F]) when the concentration of dissolved oxygen decreases as the boiling point is approached.[10] It follows that under such conditions, the corrosion rate should attain a maximum at 80°C.

The failure of the brazed fillets maintained at 60°C before those maintained at 80°C is indicative of a more complex situation than that predicted by simple rate laws. Obligately thermophilic bacteria in the failed joints appear to be implicated in the failure.

The thermophilic bacterium isolated from the failed brazed fillet forms filaments of variable lengths from 20 to over 200 µm (Figure 12.43). The optimal temperature for growth is 60° to 72°C (140° to 162° F), and the organism is completely inactive below 50°C (122°F). It grows very rapidly in the culture medium used in these experiments (doubling time less than 30 min at 65°C [149°F]). Microscopic examination of nickel

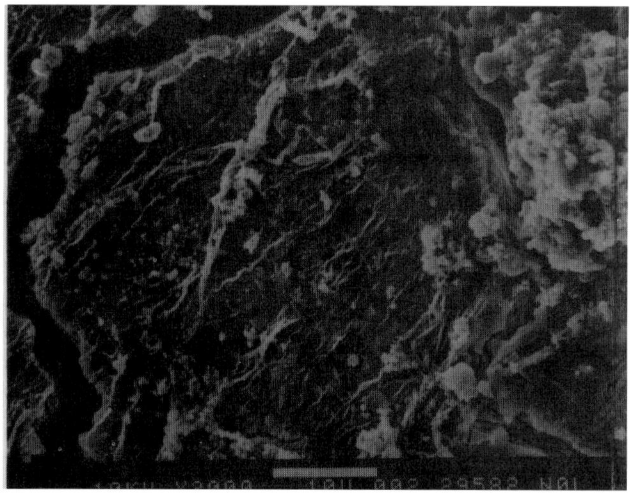

FIGURE 12.43
Thermophilic bacteria blanketing failed braze fillet.

discs stained for 3 min with 0.1% acridine orange, a fluorescent nucleic acid dye,[11] revealed attached bacterial populations of over 2 x 10^6 cells cm^{-2} after a 6-h incubation time.

Nickel electrodes were coupled using the previously described corrosion measuring device. In the absence of microorganisms, the current flow between the two electrodes was zero. After 20 h, the current at the inoculated electrode began to increase anodically and finally stabilized at 3.8 μA cm^{-2} (Figure 12.44). For comparison, an aluminum electrode galvanically coupled to a copper specimen in an identical test medium resulted in an anodic current of 80 μA cm^{-2}. At the conclusion of

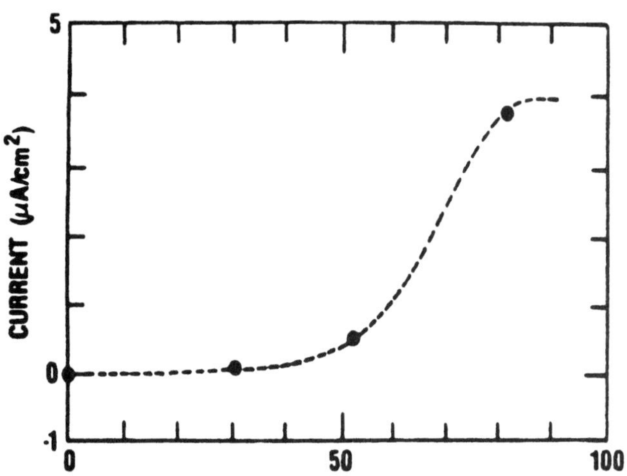

FIGURE 12.44
Corrosion current vs time resulting from colonization of the electrode by *thermophilic* species.

the corrosion measuring experiment, the working electrode was prepared for SEM. Approximately three-fourths of the surface area of the electrode was colonized by the filamentous microorganisms.

It is convenient to express the corrosion rate as milli-inches per year (mpy) (0.025 mm/y) to provide an indication of penetration. Using the traditional relationships,[12] the corrosion rate that can be attributed to the presence of the attached thermophilic bacteria after 20 h was calculated to be ~1.6 mpy (0.04 mm/y). Although the 1.6 mpy (0.04 mm/y) corrosion rate appears to be much lower than that found in the original failed system [1.9 mpd (0.0483 mm/day)], it should be remembered that these experiments were conducted in closed systems for only brief periods. Under flow-through conditions of longer duration, higher current would be expected. It is also possible that the braze material is more vulnerable to microbial attack than the solid nickel electrodes used in the corrosion measuring experiments. Two mechanisms are proposed for the observed impact: (1) formation of differential aeration cells created by respiring colonies and (2) *in situ* production of acidic metabolites.

Oxygen-deficient and even oxygen-free conditions occur beneath biofilms when oxygen is taken up by respiring microorganisms as rapidly as it diffuses to the surface. This situation creates potential differences with subsequent corrosion currents. When oxygen gas was bubbled through both cells of the experimental apparatus, the corrosion current remained at zero. Similarly, the current remained at zero during differential aeration at 23°C (73°F). However, at 60°C (140°F), the electrode in the compartment depleted of oxygen became the anode with a current of 35 μA cm^{-2} (Figure 12.45).

Most heterotrophic bacteria secrete organic acids during the fermentation of organic substrates. The types and amounts of acids produced in nature depend on the kinds of organisms present and the substrate molecules available. Organic acids from a wide range of bacteria have been shown to enhance the corrosion of a number of different types of metal.[13-16] Nickel forms a passivating film in slightly alkaline solutions. Acidic metabolites secreted by microorganisms can prevent passivation or destroy an existing passivation film.

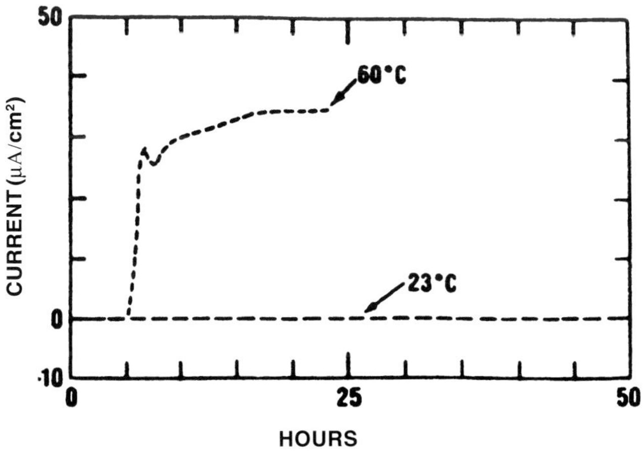

FIGURE 12.45
Corrosion current vs time resulting from abiotically created differential aeration.

Analyses indicate that two of the metabolic products of the bacterium discussed in this paper are isobutyric and isovaleric acids. When grown for 24 h in 0.1% tryptone-yeast extract medium, the bulk concentrations of these two acids were found to be 0.033 and 0.035 mM, respectively. These acids were evaluated abiotically for their contribution to the overall corrosion current at room temperature and 60°C (Figures 12.46 and 12.47). At room temperature, the concentrations

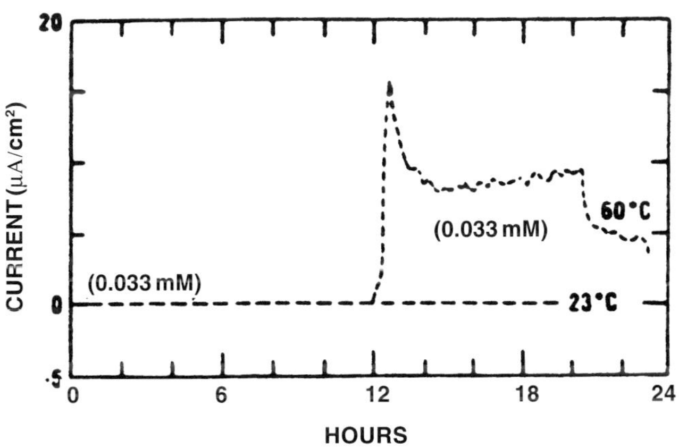

FIGURE 12.46
Corrosion current vs time resulting from two concentrations of isobutyric acid at 23 and at 60°C.

of the acids found in the bulk solution had no impact on corrosion, i.e, there was zero current. The same was true when the concentrations were increased by an order of magnitude (Figures 12.46 and 12.47).

However, at 60°C, the electrodes exposed to 0.33 and 0.35 mM concentrations of isobutyric and isovaleric acids became the anodes, and the resulting currents were in the 3.0 to 5.0 μA cm^{-2} range (Figures 12.46 and 12.47). At the higher temperature, activation energy barriers were overcome, and the acids were effective in accelerating corrosion. It is recognized that at the site of production, under the biofilm, the

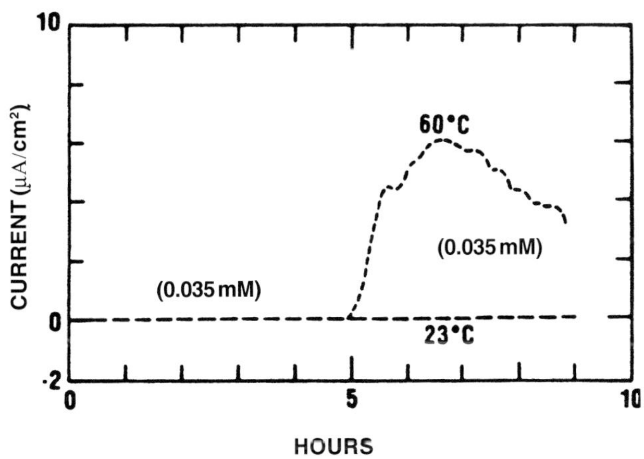

FIGURE 12.47
Corrosion current vs time resulting from two concentrations of isovaleric acid at 23 and at 60°C.

concentrations of acidic metabolites are undoubtedly much greater, and their effect is amplified. The pH value of the growth medium from which the acids were isolated dropped from 7.6 to 7.0.

Conclusion

The failure of a nickel 201 "T" brazed joint maintained at 60°C was completely unexpected and cannot be explained in terms of simple rate laws. The presence of thermophilic bacteria in the failed brazed joint implicates these organisms in the failure. The corrosion measurements confirm that the bacteria are responsible for an anodic corrosion current. Furthermore, it has been demonstrated that respiration and metabolic activities may be responsible for corrosion. The practical significance of bacteria growing in manmade hot water systems in the presence of low concentrations of nutrients has never been adequately explored.

The demonstration of a connection between these organisms and a corrosion-induced failure has enormous practical significance to the design and maintenance of hot water systems, including heat exchangers, condensers, and recirculating heaters. Apart from their ability to reproduce at high temperatures, thermophilic microorganisms resemble their mesophilic counterparts in terms of growth requirements, metabolic pathways, energy sources, and oxygen tolerance.[17-20] They are also commonly isolated from a wide variety of environmental conditions.

Acknowledgments

This work was supported in part by NORDA Program Element 61153N, H. Eppert, program manager, NORDA contribution No. 333:024:85, and ONR Contract No. N00014-81-K-0624 to Harvard University.

References

1. J.D.A. Miller, A.K. Tiller, *Microbial Aspects of Metallurgy* (New York, NY: American Elsevier Publishing Co., 1970).
2. W.P. Iverson, *Microbial Iron Metabolism* (New York, NY: Academic Press, 1974).
3. J.W. McCoy, *Microbiology of Cooling Water* (New York, NY: Chemical Publishing Co., 1980).
4. R.E. Tatnall, *Materials Performance* 20, 8(1981): p. 41.
5. D.H. Pope, D.J. Duquette, A.H. Johannes, P.C. Wayner, *Materials Performance* 23, 4(1984): p. 14.
6. C. Wagner, W. Traud, Z. *Elektrochem.* 44(1938): p. 391.
7. *Anaerobe Laboratory Manual*, ed. L.V. Holdeman, E.P. Cato, W.E.C. Moore (Blacksburg, VA: Virginia Polytechnic Institute and State University, 1977).
8. L.L. Shreir, *Corrosion*, vol. 1 (Boston, MA: Newnes-Butterworths, 1976): p. 213.
9. B. E. Conway, E.M. Beatty, P.A.D. Demaine, *Electrodhim. Acta* 7(1962): p. 39.
10. F. Speller, *Corrosion* (London, UK: McGraw-Hill Book Co., 1951).
11. J.E. Hobbie, R.J. Daley, S. Jasper, *Appl. Environ. Microbiol.* 33(1977): p. 1225.
12. EG&G Application Note 149, EG&G Princeton Applied Research (Princeton, NJ: 1978).
13. E.E. Staffeldt, O.T. Calderon, *Develop. Ind. Microbiol.* 8(1967) p. 321.
14. J.M. Burnes, E.E. Staffeldt, O.H. Calderon, *Develop. Ind. Microbiol.* 8(1967): p. 327.
15. S.A. Ashton, J.D.A. Miller, R.A. King, *Brit. Corros. J.* 8(1973): p. 185.
16. C.A. Webb, Ph.D. Dissertation (Miami, FL: University of Miami, 1975).
17. E.R.H. Gaughram, *Bacteriol. Rev.* 11(1947): p. 189.
18. T.D. Brock, *SCI* 158(1967): p. 1,012.
19. T.D. Brock, G.K. Darland, *SCI* 169(1970): p. 1,316.
20. L.G. Ljungdahl, D. Sherod, *Extreme Environments: Mechanisms of Microbial Adaptation*, ed. M.R. Heinrich (New York, NY: Academic Press, 1976): p. 147.

Crevice Corrosion of Stainless Steel in Recirculating Cooling Water[8]

By R.E. Tatnall

This case involved MIC of SS in a large cooling tower system fed by the Cape Fear River in eastern North Carolina. The plant was built in the early 1970s. Because of concern about the environment, and because the cooling tower blowdown would be a major portion of plant discharge to the river, it was decided not to use corrosion inhibitors, but rather to fabricate all water handling equipment of corrosion resistant materials. AISI 304 SS was specified for most coolers and condensers.

Cooling tower makeup was raw river water high in suspended and dissolved solids and iron. Because the plant was downstream from a large paper mill, the water also was high in chlorides and microorganisms, such as iron bacteria and sulfate reducers.

Cooling tower chemical treatment consisted of continuous chlorination to a free residual content of 0.5 to 1 mg/L, addition of caustic as required to maintain pH between 6.5 and 7.5, continuous feed of a tannin dispersant to keep silt from settling inside equipment, and shock treatment every 48 hours with a nonoxidizing biocide—quaternary amine with tris tributyl tin oxide. Chloride levels in the towers were typically about 100 mg/L and were controlled not to exceed 200 mg/L.

Approximately three years after plant startup, several pieces of SS equipment were opened for inspection. Severe crevice corrosion was found in and around gasketed joints. The corrosion sites typically showed localized soft deposits either directly over the corrosion or adjacent to it at the edge of the crevice in the gasketed joint (Figure 12.48, p. 206). These sites were generally low flow areas. The bulk deposits were usually light to medium brown and slimy to the touch. Corrosion sites directly under deposits (Figure 12.49, p. 206) were bright and active looking when the deposits were washed away. Adjacent corroded areas under gaskets, however, typically contained black corrosion products that gave off hydrogen sulfide when treated with hydrochloric acid (Figure 12.50, p. 207).

Chemical analysis of the soft brown deposits showed high levels of iron and silt, but little or no chloride, manganese, or sulfur. As mentioned, both suspended iron and silt were present in the water at high levels. Bacteria analysis showed the bulky deposits to have high counts of *Pseudomonas* slime-forming bacteria and a few *Sphaerotilus* filamentous iron bacteria. *Desulfovibrio* SRB were confirmed in the black underlayer.

An important observation is that, although numerous cases of this type of corrosion were seen at this site, no corrosion was ever seen in the absence of bulky deposits. The proposed mechanism for this corrosion is:

- Slime formers, which are free floating in the system, attach and multiply at low velocity sites, such as gasket joints and other crevices.
- As the slime deposit grows, it traps suspended solids from the water, including iron oxides and free-floating filamentous bacteria, which thrive in the iron rich deposits.
- The slime formers and iron bacteria

[8]Originally published as CORROSION/81 paper no. 130 (Houston, TX: NACE, 1981)

consume any oxygen that diffuses into the deposit, thus forming an exaggerated anaerobic condition underneath.

- SRB find their way into the deposits and concentrate in the anaerobic zone at the metal surface.
- The combination of the oxygen-free condition and the sulfide generated by the SRB destroys the passivity of the SS, leading to localized corrosion.

It is easier to diagnose the problem than cure it. Even though bacteria are present in the water source, it is surprising that they survive the combination of chlorine and nonoxidizing biocide. A program is under way to identify and put into practice a more effective chemical treatment. In the meantime, temporary equipment repairs are being made as follows:

- Physically remove deposits and clean the equipment;
- Weld overlay the corroded areas and resurface as required. AISI 316 SS weld filler metal is used to provide better corrosion resistance; and
- Where weld overlay is impractical, corroded areas are blast cleaned and filled with an epoxy putty to reseal the gaskets.

A program is also under way to evaluate 90/10 cupro/nickel and AISI 316 SS as alternative materials of construction if biological control proves impractical.

Penetration of Stainless Steel Following Hydrostatic Test [9]

By J.G. Stoecker

In 1978, a major expansion project at an East Coast plant site required a large amount of 7.5 cm and 20 cm (3 in. and 8 in.). Schedule 10 AISI 304 SS piping. The system was completed in the early fall and hydrostatically tested with potable well water. This water, which was treated by continuous chlorination, typically analyzed about 250 mg/L total chloride. After testing, the lines were drained and steam traced. The system sat idle about five months.

In late March 1979, the 7.5 cm (3 in.) section was placed in service. Within less than 24 hours, leaks were noticed in the piping butt welds, and the system was shut down. Another hydrostatic test was run, and leaks were confirmed at welds—not only in the smaller diameter piping, but in the unused larger lines as well. Spot radiography indicated that 80% of all welds contained large subsurface cavities, such as those shown in Figure 12.51.

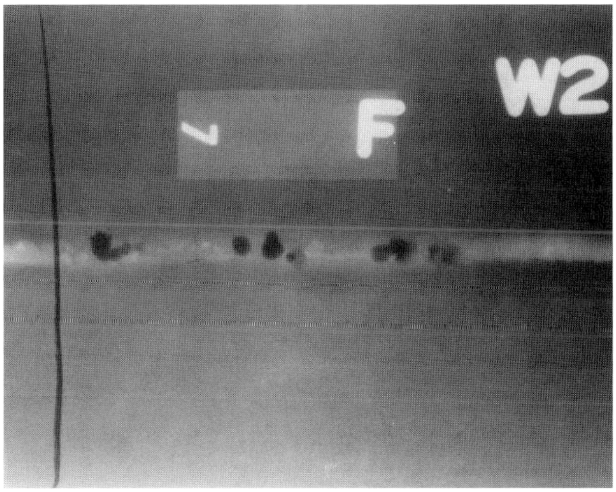

FIGURE 12.51
Radiograph of weld seam showing subsurface cavities, not visible from either the inside or outside surfaces.

Inspection of the inside surfaces showed a powdery, reddish-brown deposit along the weld seams (Figure 12.52, p. 207). A cross section through one typical cavity is shown in Figure 12.53. While it is likely that a small amount of water remained in the lines after draining, this void, as with many others, was well up on the sidewall of the horizontal line, and thus definitely not submerged in any residual water.

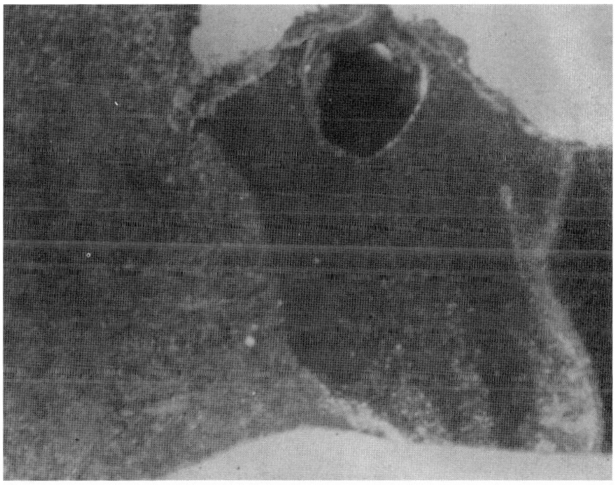

FIGURE 12.53
Weld cross-section showing typical subsurface cavity. In this case, corrosion was limited entirely to the weld metal. Note minute opening to the ID surface. (~ 15X)

This case is similar to one reported by Kobrin in 1976, where *Gallionella* iron bacteria were confirmed to have caused the corrosion.[1] In that case, the deposits were found to be high in iron and manganese, apparently concentrated by the

[9]Originally published in *Materials Performance* 20, 8(1981): p. 43.

Gallionella and oxidized to their higher valence states. For reasons not totally clear, the deposits also contained high chloride concentrations. The resulting ferric and manganic chloride environment is known to produce subsurface cavities, such as those we found.

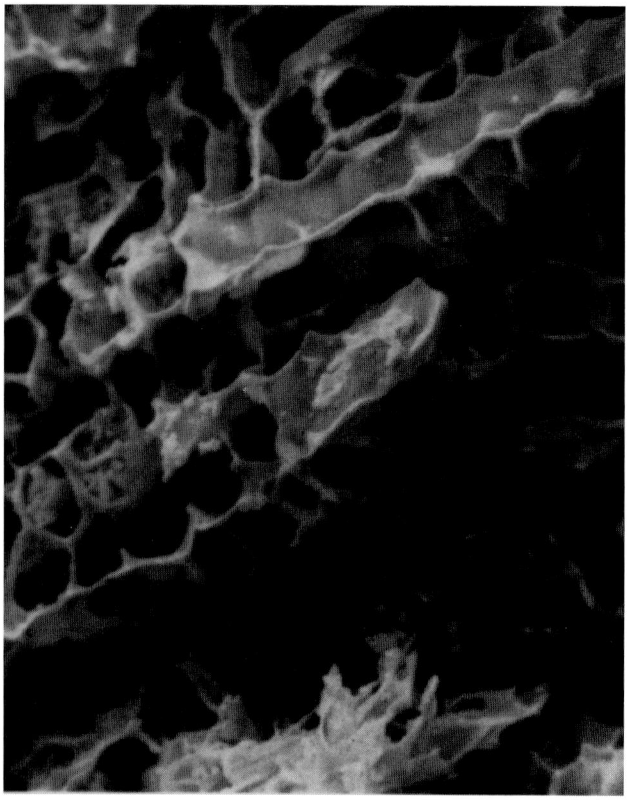

FIGURE 12.54
SEM micrograph showing matrix remaining after preferential corrosion of the ferrite phase. (~1000X)

Kobrin reported preferential corrosion of the delta ferrite structure in the weld metal. We found the same thing, shown in the scanning electron micrograph in Figure 12.54. This preferential attack of the ferrite phase is typical of corrosion by ferric chloride and dilute HCl.

It is not understood why *Gallionella* colonies on SS are almost always within or adjacent to welds. Our welds showed a variety of minor defects such as porosity, slag inclusions, and joint misalignment, but the corrosion sites did not coincide with any such defects. In fact, *Gallionella* corrosion has been observed on apparently defect-free welds that had been ground smooth and flush; adjacent to small particles of weld spatter on tank walls and floors; and even opposite external welds that had not melted through to the affected surface—but almost always associated with welds in some way.

During the inspection, widespread stress corrosion cracking (SCC) was observed, apparently due to the concentration of chlorides by evaporation of the residual water with heat provided by the steam tracing. The piping was not repaired, but was replaced in kind and pneumatically tested.

Reference

1. G. Kobrin, *Materials Performance* 15, 7(1976): p. 38.

Massive Utility Condenser Failure Caused by Sulfide-Producing Bacteria[10]

By P.R. Puckorius

The examination of perforated SS condenser tubes in a large utility condenser that failed by pitting during a six-week shutdown period is described. A successful temporary means of repair that enabled the pitted tubes to remain in service is described.

Introduction

This paper reports an extensive study to determine the cause of an unexpected massive failure of a utility power condenser from the water side. This condenser, composed of AISI 304 SS tubes, was 90 ft (27.4 m) long and contained slightly more than 28,000 tubes. It was a single-pass condenser and was designed to drain out completely when shut down. This station is a 900 megawatt (MW) power plant. The water used for cooling comes from a moderate lake of about 1,400 acres.

Because of a coal plant explosion, the cooling system was shut down and remained out of service for approximately six weeks. It was then cleaned by water blasting and tested to determine its integrity. The investigation revealed that about 3% to 5% of the tubes were leaking slightly. These tubes were plugged, and the unit was put on line, which led to additional leakage. The system was shut down and tested again. Additional leaks were subsequently uncovered. When a vacuum was established on the steam side, the popping of tubes was heard from the water side while an examination was being conducted. From all indications, the tubes had been perforated as well as attacked since only a thin film of metal remained. This thin film failed under a light vacuum load during the testing procedure.

Examination of Corroded Tubes

Several tubes were removed and examined to determine the location and type of failures. Most of the tube failures had occurred on the bottom half of tubes in the first 4 ft (1.22 m) from the inlet by intense pitting. Many cases showed perforation of the tubes as well as partial penetration.

The study was initiated with an examination of the waterbox tubes and a detailed analysis of deposits. Included was a review of water quality and an extensive examination of the failed SS tubing. The general conclusions were that the failure had occurred because of biological organisms, specifically the activity of sulfide-producing bacteria.

The condenser and the cooling water system were examined first. Figure 12.55 (p. 208) shows the type of fish, mud, and biological organisms found within the condenser when the unit was shut down after being restarted when repairs were completed. The presence of substantial fish, mud, and biological organisms was a direct result of deteriorated screens on the intake, thus causing considerable biological contamination.

Figures 12.56 (p. 208) and 12.57 show the presence of biological slime and portions of fish removed from the tubes during the cleaning process. Figure 12.57 shows the presence of slimy deposits within the tubes as well as the

[10]Originally published as CORROSION/83 paper no. 248 (Houston, TX: NACE, 1983)

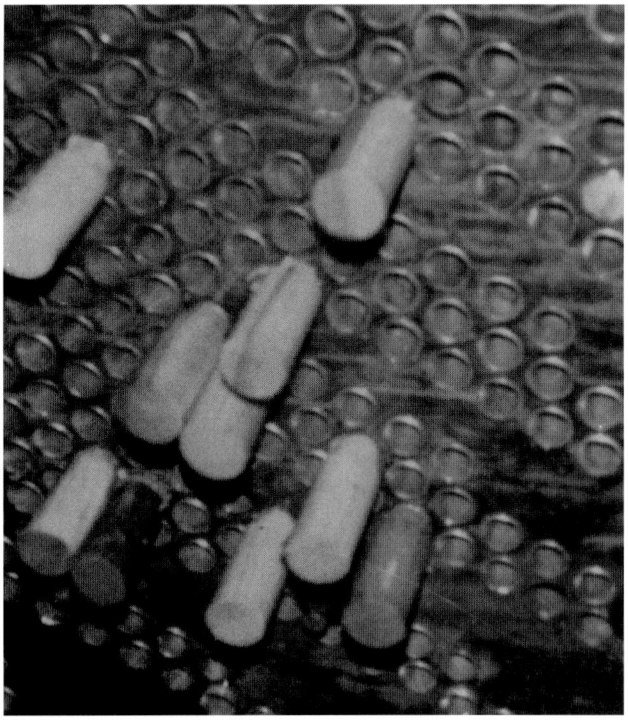

FIGURE 12.57
Slime, mud, and fish in partially cleaned tubes.

extensive plugging accomplished with both purchasedplugs and hardwood plugs prepared by the utility's carpenters. It was found that the first several feet of tubing did not completely drain of water. It was not known whether the condenser was originally capable of complete drainage or not.

Figure 12.58 shows the hang-up of water in the bottom of the tubes. This enabled biological organisms to grow extensively, aided by the presence of moisture, nutrient, and lack of effective biological control measures.

The past water treatment program had used only sodium hypochlorite at a level that would normally control most biological organisms. Because of the increase of deposits and nutrients from decaying fish, however, the chlorination program may not have been sufficient to effectively control all types of organisms. This was because of the high chlorine demand of the biological and marine deposits.

Samples of the slimy deposits were submitted to a laboratory for identification of all possible organisms. Specific searches were made for iron bacteria and the corrosive bacteria frequently referred to as sulfate reducing bacteria (SRB). The presence of SRB, in addition to slime-forming bacteria, algae, and considerable decomposed organic matter, was verified.

A number of tube specimens were obtained and examined both visually and microscopically. Figure 12.59 (p. 209) illustrates the type of pitting that had occurred. Some tubes showed many pits with perforations and extensive depth. Most tubes showed an iron stain around pitted areas. A metallographic examination indicated the alloy microstructure was typical of annealed 304 SS with no carbide precipitation. The pits through the wall were smooth-sided with intergranular corrosion.

Attention was then focused on the deposits and type of corrosion occurring on the tubes. In a number of tubes that had not been mechanically cleaned, a peculiar type of deposit was noted, as illustrated in Figure 12.60 (p. 209). Small volcano-like deposits were noted with black deposits around the edge or covering over the volcano area. From all indications, pitting had started in these areas. Figures 12.61 and 12.62 show the nature of the attack in more detail. Deposit samples of the tube surface at the center and edges of the pit were analyzed. In all cases, no chlorides were found, although the presence of silica and calcium was at times noted, along with some aluminum and sulfur.

The presence of the sulfur strongly suggested

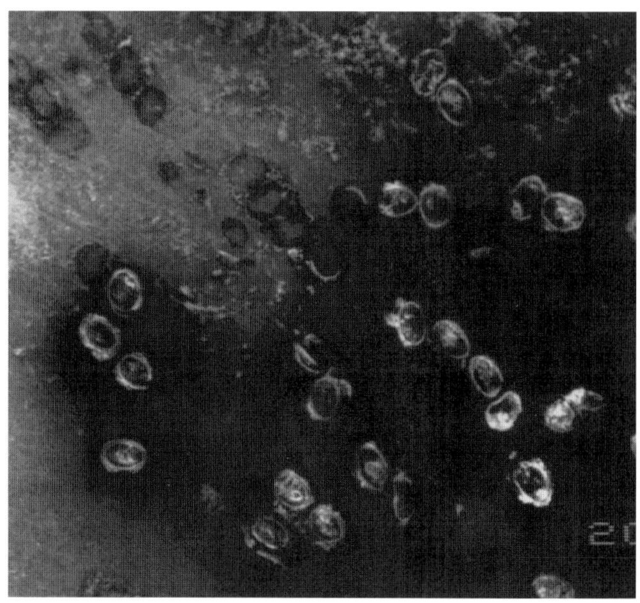

FIGURE 12.61
Magnification of black volcano-type deposits due to sulfide-producing bacteria.

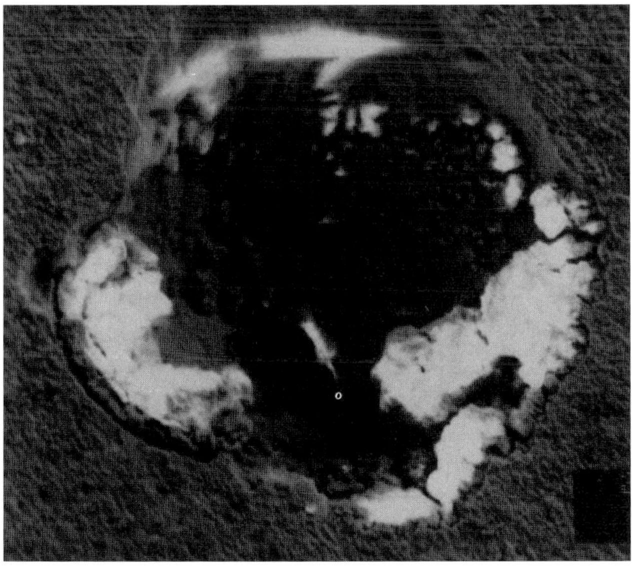

FIGURE 12.62
Pit and surface deposits. White, bright deposits are deposits within pit.

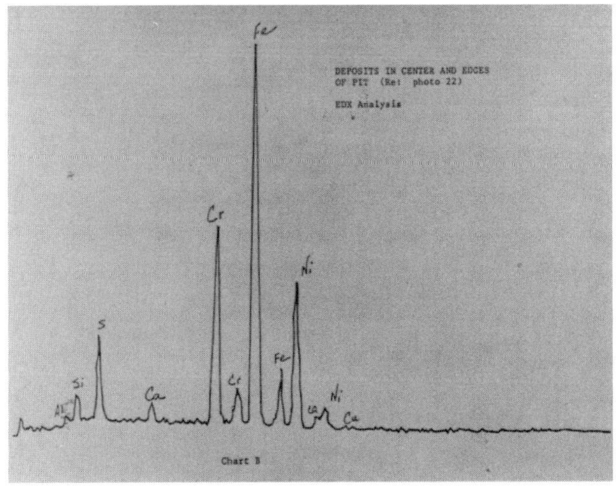

FIGURE 12.64
Deposits in center and edges of pit. EDX analysis.

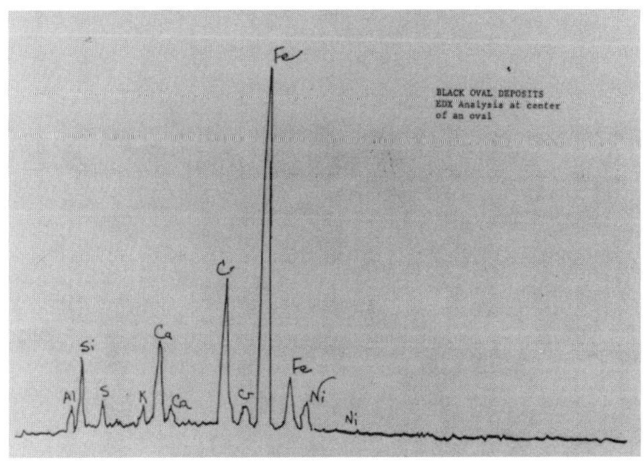

FIGURE 12.65
Black oval deposits. EDX analysis at center of oval.

SRB. These organisms are known to severely attack mild steel and can cause a reducing atmosphere because of their production of hydrogen sulfide along with a low pH condition that can very likely deteriorate the oxide film of SS.

The EDX analyses are shown in Figures 12.63 through 12.66. As seen in Figures 12.65 and 12.66, both the center and the edge of the ring show the presence of sulfur and no chlorides in the black oval deposits. Figure 12.64 shows substantially more sulfur at the center and edges of the pit. During the examination, a musty odor and

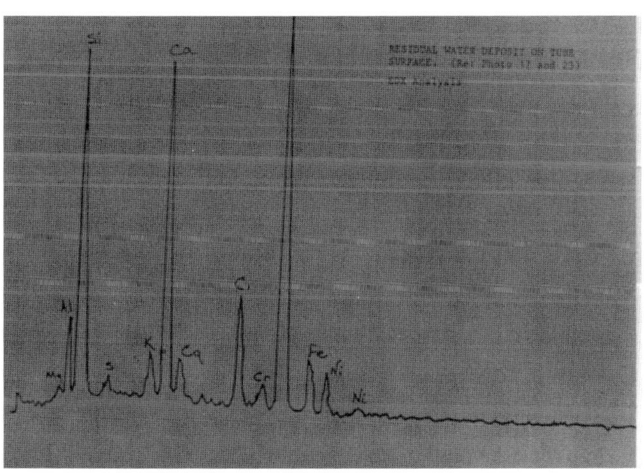

FIGURE 12.63
Residual water deposit on tube surface. EDX analysis.

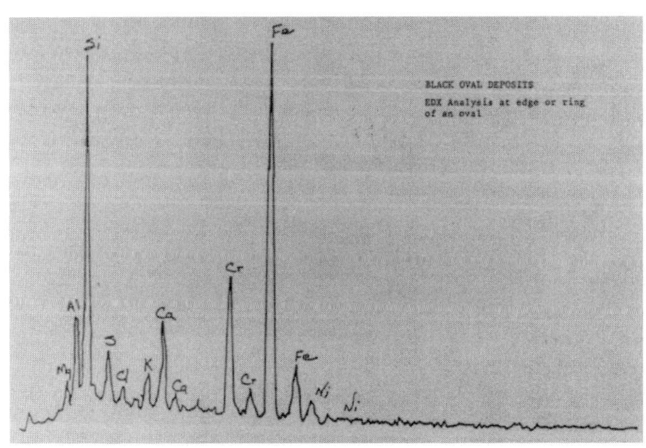

FIGURE 12.66
Black oval deposits. EDX analysis at edge or ring of an oval.

the smell of hydrogen sulfide were noticed.

The analyses confirmed the presence of sulfide-producing bacteria. Due to the holdup of water in the condenser, and with the presence of considerable amounts of nutrient, these organisms continued to multiply rapidly during the outage and prior to any cleaning. This caused the perforation of the tubes within eight weeks.

Repair Procedure

The recommended corrective action was to clean the tubes and sterilize them with a specifically known biological chemical that is effective for controlling sulfide producing bacteria. Upon sterilizing the condenser, all efforts were made to return the unit to service by plugging the failed tubes. A unique method was selected to plug the pits, but not the tubes. This technique would enable the condenser to return to service within a couple of weeks. Replacement of tubes could not be done for at least 12 to 18 months. Plugging of tubes would require 10,000 to 12,000 plugs (one for each end of the tube) and complete loss of the tube for heat transfer purposes.

This unique procedure was recommended by George Moller, who had considerable experience with this technique in the petroleum and petrochemical industries.

Fortunately, the pits in the tubes occurred in the first 4 ft from the inlet and primarily in the bottom of the tube. Thus, the technique was ideally suited for this utility power condenser.

Essentially, the method involves cleaning the tubes, then applying a quick-set epoxy sprayed into the tubes to a distance of 5 to 6 ft. This epoxy sets, hardens, and seals the perforations. It also fills the as-yet unperforated pits and prevents them from developing further. This method was used quite successfully, and the unit returned to service. It has been operating at 100% efficiency for 12 months.

This technique is only temporarily effective. In the refinery and petrochemical industries, such repairs have lasted in excess of five years in a number of instances.

The remaining consideration was to maintain good water treatment practices as well as proper maintenance of the screens of the pump suction. Chlorination practices were increased to maintain good control of biological organisms. Every effort was made to keep the surfaces clean by modifying the water composition and water treatment to eliminate mud, silt, and scale where organisms might accumulate.

Summary

The perforation of AISI 304 SS condenser tubes that occurred during a six week shutdown period was shown to be caused by failure of the tubes to drain completely and by the presence of sulfide producing bacteria, plus an ample supply of nutrients in the form of fish and other biological material. To avoid loss of heat exchange surface that would occur if all the pitted tubes had to be plugged, a novel means of repair using an epoxy coating was found to be successful for a 12-month period.

Acknowledgments

We acknowledge the excellent cooperation provided by George Moller, his detailed analytical work with the electron microscope and other metallurgical techniques, as well as the personnel at the utility power station for their cooperation in determining the cause of this failure.

Waterside Penetration of Ferritic-Austenitic Stainless Steel Alloy 255 Heat-Exchanger Tubes[11]

By J.G. Stoecker

This case deals with the localized corrosion and penetration of 19.1 mm (0.75 in.) OD x 1.9 mm (0.075 in.) wall ferritic-austenitic SS alloy 255 tubes in a saltwater heat exchanger. The saltwater is on the tube side of the exchanger and is drawn from Galveston Bay. Inlet water temperatures vary from 15 to 29°C (60 to 85°F) depending on the time of year. Outlet temperature is controlled at 42°C (108°F). The water velocity in the tube side of the exchanger ranges from stagnant in the heads and outermost tubes to a fairly high and constant velocity in the center of the tube sheet and in the centermost tubes. Bennett reports that microbiological growth increases when flow is less than 0.76 m/s (2.5 ft/s).[1]

Microbiological analyses of the saltwater in and out of this exchanger were conducted for samples collected on four consecutive days. The results are shown in Table 12.14. Unfortunately, these analyses were conducted five to seven days after the samples were taken, and the data are of limited value. This document is quite impressive, but proves only that some of the examples did contain bacteria that somehow survived this extended waiting period.

The bacteria counts in the samples cannot be used as an indicator of activity in the flowing dynamic stream from which the samples were taken. Bacterial counts must be made within 24 hours or less after sampling, coupled with special sample-handling procedures, for the analysis to accurately reflect microbial activity in the stream.

The corrosive penetration of the tube from the inside or saltwater side as shown in Figure 12.67 could be pitting typical of chemical attack. Some of the pits did, however, exhibit some unusual characteristics in these otherwise clean tubes.

Figure 12.68 shows an example of a pit with a reddish-brown deposit immediately downstream of the localized attack. Unfortunately, this bright-colored deposit is not evident in the black and white photograph. The reddish-brown deposit in the vicinity of corrosive attack by a delta ferrite-containing alloy has been associated previously by the author with *Gallionella* activity.[2] The delayed microbiological analysis, however, failed to identify this species in any of the saltwater samples tested. MIC is indicated, but cannot be proven definitively.

References

1. E. Bennett, *Lubrication Engineering* July (1972).
2. R. Tatnall et al., *Materials Performance* 19, 8(1981): p. 41.

[11] Originally published as CORROSION/83 paper no. 245 (Houston, TX: NACE, 1983).

TABLE 12.14
Microbiological Analysis of Saltwater In and Out of Exchanger [1]

	1st Day		2nd Day		3rd Day		4th Day	
	In	**Out**	**In**	**Out**	**In**	**Out**	**In**	**Out**
BACTERIA								
Anaerobic slime-forming								
Nonspore forming								
Flavobacterium	Neg 1/1,000	Neg 1/1,000	Neg 1/1,000	Neg 1/1,000	Neg 1/1,000	Neg 1/1,000	Neg 1/1,000	Neg 1/1,000
Mucoids	Neg 1/1,000	Neg 1/1,000	Neg 1/1,000	Neg 1/1,000	Neg 1/1,000	Neg 1/1,000	Neg 1/1,000	Neg 1/1,000
Aerobacter	Neg 1/10	Neg 1/10	Neg 1/10	Neg 1/10	Neg 1/10	Neg 1/10	Neg 1/10	Neg 1/10
Pseudomonas	2,000	Neg 1/1,000	1,000	Neg 1/1,000	Neg 1/1,000	Neg 1/1,000	Neg 1/1,000	Neg 1/1,000
Spore-forming								
B. subtilis								
B. cereus								
B. megatherium								
B. mycoides								
Anaerobic Corrosive								
Desulfovibrio	Neg 1/10	20	Neg 1/10	Neg 1/10	Neg 1/10	Neg 1/10	10	Neg 1/10
Clostridia								
Iron-depositing								
Sphaerotilus	None	None	None	None	None	None	None	None
Gallionella	None	None	None	None	None	None	None	None
Other bacteria	None	None	None	None	None	None	None	None
Total count	Neg 79,000	1/1,000	Neg 13,000	Neg 1/1,000	Neg 1/1,000	Neg 1/1,000	Neg 1/1,000	1/1,000
FUNGI								
Molds								
Aspergillus *Penicillium* *Trichoderma* *Alternaria*	None	None	None	None	None	None	None	None
Yeasts								
Torula *Monilia* *Saccharomyces* *Rhodotorula*	None	None	None	None	None	None	None	None
ALGAE								
Blue-green								
Oscillatoria	None	None	None	None	None	None	None	None
Green								
Chloroccus	None	None	None	None	None	None	None	None
Diatoms	None	None	None	None	None	None	None	None
Other organisms	None	None	None	None	None	None	None	None
Chlorine residual		0.78		0.35		0.17		0.14

[1] Physical appearance of all samples: liquid with floc. All counts express quantity of organisms per mL of sample.

Study of Biological Corrosion in High-Temperature Demineralized Water[12]

By J.G. Stoecker and D.H. Pope

Introduction

The phenomenon of MIC is now generally accepted as being one of a number of mechanisms whereby corrosion of a variety of metallic materials can occur. Acceptance of this phenomenon as a scientifically valid topic for investigators has been brought about by the activities of several investigators in both engineering and scientific disciplines. The reader is referred to a number of recent publications dealing with case histories, methods of detection, and treatment for further details on these aspects of MIC.[1-5] Recent reviews and guides to MIC by the authors and several colleagues are now available.[5,6] In addition, several older reviews of aspects of MIC problems are valuable resource documents.[7-10]

The details of the multidisciplinary study of a case history of MIC are presented to add to the growing reservoir of knowledge on the subject in hopes that solutions to the problems will accompany knowledge. This study involves the localized corrosion of AISI 304 SS and is unique because it involves a single microbiological species at elevated temperature in relatively pure demineralized water. Both the microbiological and metallurgical aspects of the study are described. These aspects include microbiological testing and descriptions of some characteristics of the bacteria involved. Details of the metallurgical and complete electron probe microanalyses of the various cracks associated with the localized corrosion and deposits are also presented.

Background

Figure 12.69 (p. 210) shows the damaged and leaking tank during replacement in August 1984. This tank is constructed of AISI 304 SS, is 12 ft (3.7 m) in diameter and 19.3 ft (5.9 m) high. The shell is 0.5 in. (13 mm) thick, and the heads are 0.75 in. (19 mm) thick. The tank is insulated and has an internal steam heating coil for high temperature operation.

The tank was placed in hot demineralized water service with an operating temperature of 150°C (302°F) in 1976. The operating temperature was reduced to 80°C (176°F) in 1979. Control of the operating temperature since 1979 has not been exact, and the temperature of the water has been allowed to fluctuate from 75° to 90°C (167° to 194°F). Water temperatures below 75°C cause process problems and are avoided. Treated water make-up begins when the level in the tank drops to 80% and stops when it rises to 95% capacity.

The first internal inspection of this vessel was conducted in 1981. This inspection revealed the existence of numerous deposits over the entire interior surface with concentrations at the welds and stressed areas in the tank. The exterior surface of the deposits in contact with the water were light rust colored on the top surface and changed to a dark brown-black color at the tank wall. The metal underneath the deposit and in the pit in the tank was powdery and grey-green in color. The measured pit depth

[12]Originally published as CORROSION/86, paper no. 126 (Houston, TX: NACE, 1986).

TABLE 12.15
Water Analyses

Year Sample Taken	1982	1983	1984
Total magnesium hardness as $CaCO_3$, ppm	0.10	0.5	0.13
Total hardness, sum of ions, $CaCO_3$, ppm	0.15	1.0	0.28
Total copper as Cu, ppm	0.05	0.05	0.05
Total calcium hardness as $CaCO_3$, ppm	0.14	0.50	0
Sulfate and sulfite as SO_4, ppm	--	32	29
Specific conductivity, UMHOS	219	190	197
pH	6.2	5.9	6.1
Chloride as Cl, ppm	24	19	21
Alkalinity, M, as $CaCO_3$, ppm	16	5	14
Alkalinity, P, as $CaCO_3$, ppm	0	0	0
Total iron as Fe, ppm	0.05	0.06	0.05
Sodium as Na, ppm	45	41	37
Silica as SiO_2, ppm	--	13	15
Phosphate, total as PO_4, ppm	--	0.4	--
Phosphate, total inorganic, as PO_4, ppm	--	0.2	00
Phosphate, ortho, as PO_4, ppm	--	0.2	00

ranged from 1/64 to 5/64 in. (0.40 to 2.0 mm) deep. What appeared to be lime scale had fallen off the internal heating coil and was laying on the bottom of the tank. The coil exhibited areas of corrosion beneath residual scale, which appeared to be a concentration cell form of corrosion rather than the sharp pitting observed on the walls of the tank.

The recommendation was an attempt to identify the cause in order to counteract the pitting. It was determined that this type of attack, if left unchecked, could produce leaks and render the tank unusable before it had completed its expected life cycle. It was also recommended that this tank be inspected every two years.

Another internal inspection of the vessel was conducted in 1983 in keeping with the two-year frequency of inspection recommended in 1981. This inspection revealed the existence of numerous barnacle type deposits on the walls of the tank, which were concentrated mostly around the heat affected zones (HAZ) of the welds and stressed areas. The deposits were rust colored externally with a black layer at the tank wall. The pitting in the tank wall underneath the deposits was sharp, and the depth ranged from 1/8 to 3/16 in. (3.2 to 4.5 mm). The pits were filled with a powdery grey-green material.

The inspection revealed that the tank was leaking at two or more locations. The number and exact location of the leaks could not be determined from the inside of the tank because of the 300 to 400 barnacles on the surface of the tank. The number and locations of the leaks could not be determined from the exterior surface of the tank unless the insulation was removed completely, which was deemed impractical.

Repair welding the pits was not recommended at this inspection. The rationale was that welding the pits would probably result in cracking and additional pitting since the observed damage was preferential to the HAZ of the original welds (i.e., for each pit that was repair welded, three or four more pits could develop surrounding the repair weld). The recommendation was to replace the existing tank with one constructed of AISI 316 SS. The replacement recommendation was based on the pit depth being one-third of the way through the current tank wall and the existence of at least two leaks in the vessel.

Water Analyses

Early in 1984 the authors were apprised of the MIC problem and became part of the investigation. The first step in our investigations was to determine the quality of the water in the system. Table 12.15 shows typical chemical analyses for 1982, 1983, and 1984. Water treatment consists of injecting Midwestern well water with chlorine to neutralize residual ammonia and eliminate microbial growth. The water is held in a residence tank from which it is pumped through activated carbon filters and hydrogen resin filters to a degassifier tower. The pH is adjusted with caustic soda after degassification, and then the water is pumped through sodium zeolite softeners to the tank.

Microbiological analyses of the water began May 22, 1984, when plant personnel sampled the feed stream to the tank. Samples were collected into 50-mL sterile plastic centrifuge tubes containing INT dye to provide a final concentration of 0.02% and a mixture of organic respiratory substrates from Biotest, Inc. These were incubated at room temperature for 60 min. Buffered formalin was added to give a final concentration of 2% formaldehyde. These were shipped to Biotest for analysis using fluorescent isothiocyanate (FITC) staining and direct fluorescent antibody techniques for total direct microscopic counts and SRB counts, respectively.[11]

Portions of the samples were also inoculated into modified Postgate's medium B, Biotest, and Sulfate API Broth (Difco) in an attempt to grow SRB. These were incubated for 30 days before scoring them as negative.

Tank Inspection and Sampling

On August 8, 1984, the authors took part in the final inspection of the tank during its replacement. Figure 12.70 (p. 211) shows typical deposits on the interior surface of the tank. The deposits in this view are associated with the HAZ of a vertical seam weld in the tank. As noted previously, the deposits in Figure 12.70 are generally rust colored but change to dark brown-black at the metal surface. The white line in this view encloses the area of the specimen of tank wall that was removed for both microbiological and metallurgical analyses.

Microbiological Analyses

Samples of nodule materials were plated onto plate count agar and incubated at 25° and 45°C (77° and 113°F) aerobically and anaerobically in an attempt to isolate organisms.

Microscopic and culture analyses were all negative for SRB. Nodule materials were found to contain several hundred thousand to several million bacterial cells per nodule. These rods proved to be gram positive (Figure 12.71) and appeared to form both spores and slime. This suggests that they are *Bacillus sp.* All isolates grew at both 25 and 45°C. It also should be noted the same type of organisms have subsequently been isolated from nodules associated with MIC of carbon steel, other SS cases, and aluminum bronze.

FIGURE 12.71
Photomicrograph of rod-shaped bacteria. (1000X)

Metallurgical and Electron Probe Microanalyses

Figure 12.72 is a photomicrograph of a section through a typical deposit and pit in the wall

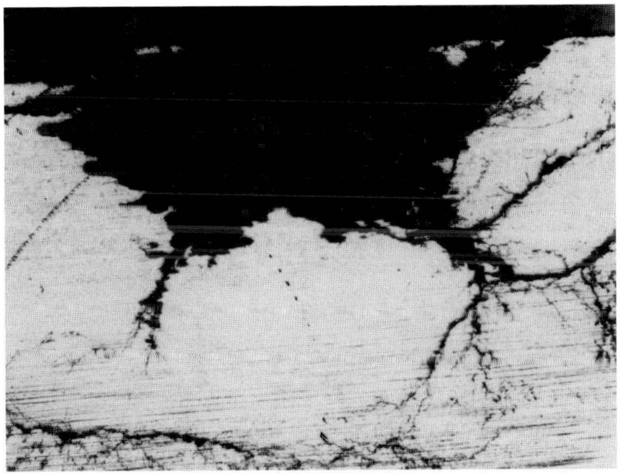

FIGURE 12.72
Photomicrograph of section through a typical deposit and pit in the wall of the tank. 10% oxalic etch. (12.8X)

of the tank. The cracks emanating from the pits explain why the tank was leaking even though the pits were measured at only one-third of the depth of the tank wall thickness. Figure 12.73 is a photomicrograph at a higher magnification of the

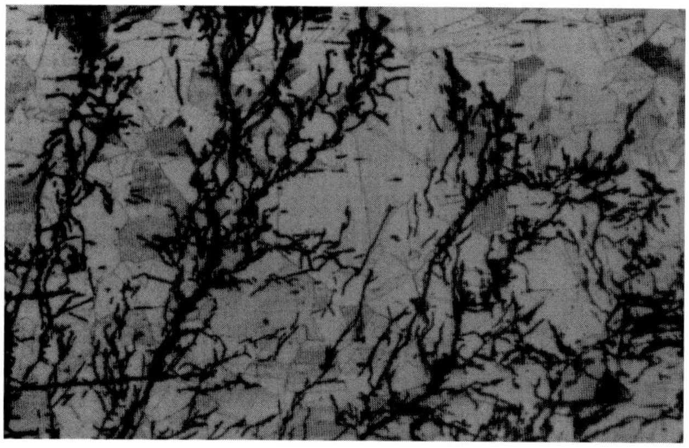

FIGURE 12.73
Photomicrograph at higher magnification of cracks shown in Figure 12.72. Note that the branched transgranular cracks in this view are typical of chloride SCC of austenitic SS. 10% oxalic etch. (125X)

cracks shown in Figure 12.72. The branched transgranular cracks in this view are typical of chloride SCC of austenitic SS.

Electron probe microanalyses (EPM) was perceived to be a key tool in this metallurgical investigation. Initial efforts in the investigation were devoted to the analyses of suspect regions or rather crudely prepared sections that had been cut from the specimen taken from the sample of tank wall (Figure 12.70, p. 211). The sections were polished using only rough grit paper in an effort to eliminate possible contamination by polishing agents and to eliminate the possible introduction of water into the areas of interest. A major concern was the effect on the analytical results that might be induced by the rough sanded conditions of the section.

The EPM technique depends on having the highly polished flat surface for analysis. Roughness should have a significant effect only on the quantitative results, and at that point in the investigation, quantification was not a major concern. Areas analyzed were all located deep within the section at the ends of cracks that originated at the surface deposit. At these points the extension cracks seemed to be grain boundary deposits.

Four different areas were analyzed. All appear to have the same basic composition (i.e., higher concentrations of sulfur and chromium within the defect area and corresponding lower levels of both iron and nickel). It is considered significant that neither oxygen nor carbon was detectable in these areas.

The corrosive attack appeared, at least initially, to be a grain boundary phenomenon. Images shown in Figures 12.74 and 12.75 in regions

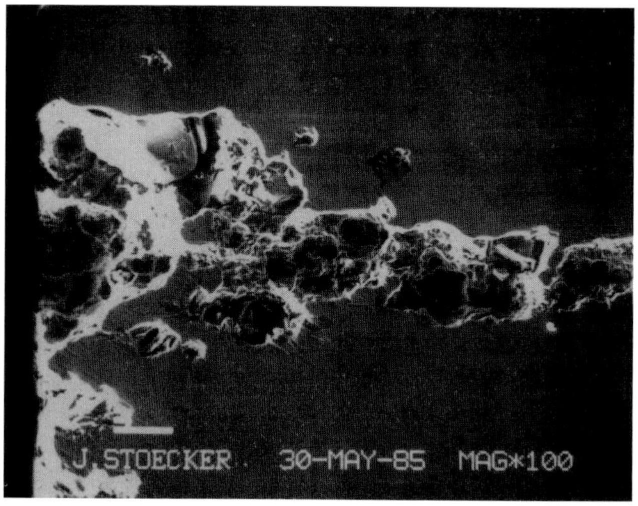

FIGURE 12.74
Photomicrograph of intergranular corrosion in the AISI 304 SS metal matrix associated with a typical pit and microbiological deposit. (100X)

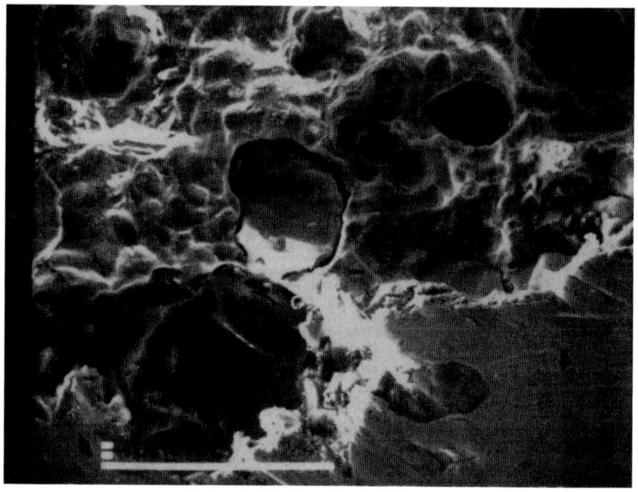

FIGURE 12.75
Photomicrograph at higher magnification of intergranular grain lifting corrosion of the same attack as shown in Figure 12.74. (400X)

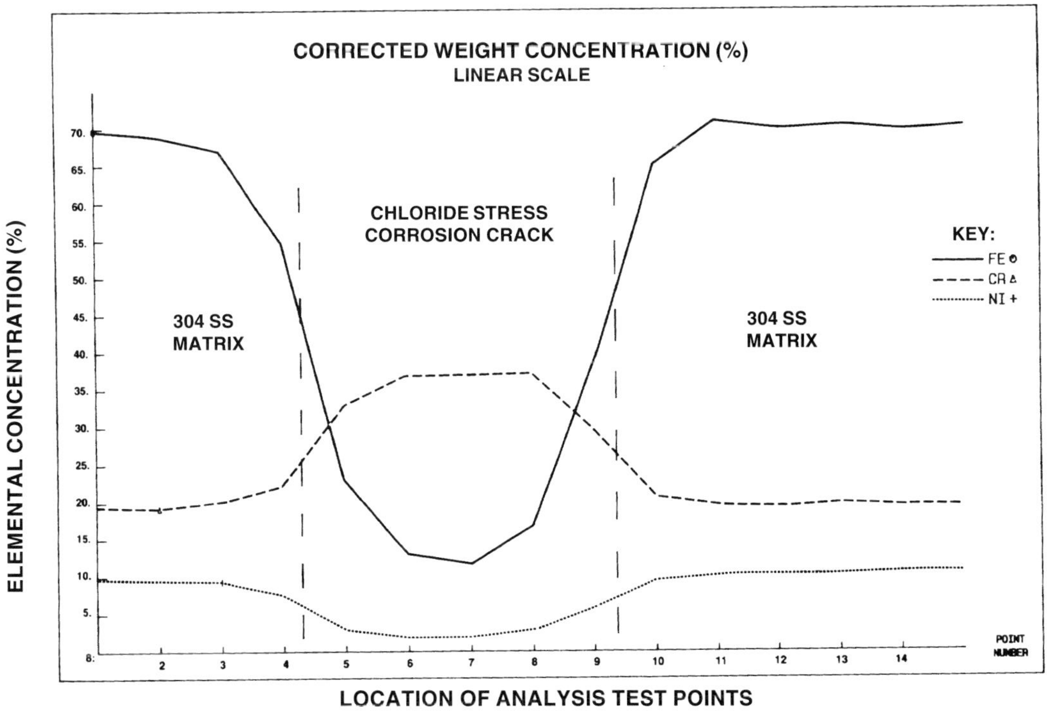

FIGURE 12.76

Electron probe microanalyses elemental plots across a typical crack in a rough sanded pitted cross-sectional specimen from the tank wall. From left or right, top line depicts iron content, middle line depicts chromium content, and bottom line depicts nickel content.

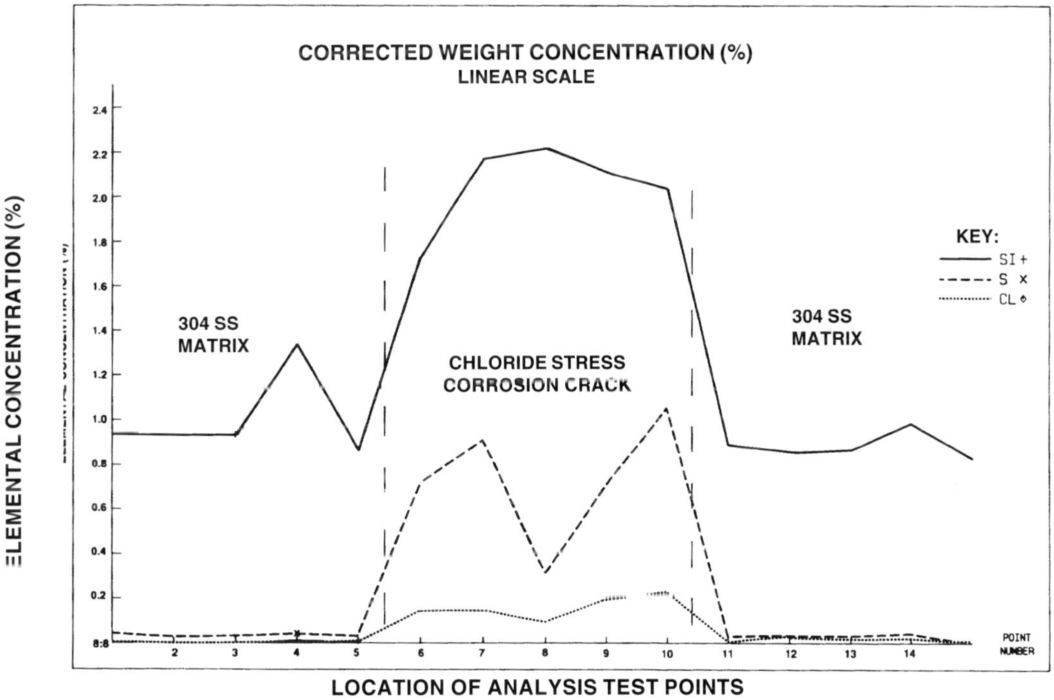

FIGURE 12.77

Electron probe microanalyses elemental plots across a typical crack in a rough sanded cross-sectional pitted specimen from the tank wall. Top line depicts silicon content, middle line depicts sulfur content and bottom line depicts chlorine content.

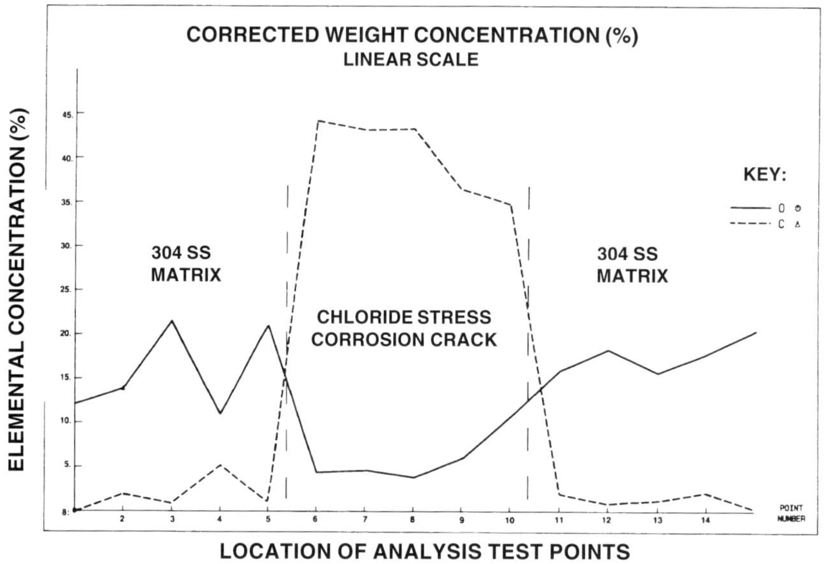

FIGURE 12.78

Electron probe microanalyses elemental plots across a typical crack in a rough sanded cross-sectional pitted specimen from the tank wall. Top line in the crack depicts oxygen content and bottom line in the crack depicts carbon content.

close to the original surface appeared to be entire exposed grains. Deeper within the cross section, where gross attack ceased, grains were observed that seemed to have been detached from the metal matrix. If that were the case, as further corrosion occurred and after sufficient amounts of the surrounding matrix were removed, entire grains would be freed to drop out of the matrix and wash from their original areas within the section. At this point, the specimen was polished further in an effort to provide suitable sections for investigation at higher resolution.

Figures 12.76, 12.77, and 12.78 are EPM plots across a typical crack in a rough sanded cross-sectional specimen from the specimen cut from the tank wall. Figure 12.76 shows the elemental plots for iron, chromium, and nickel from one AISI 304 SS matrix to the other across a crack. The top line in the figure is a plot of the iron content, which is at a high level of 70.82% in the metal matrix and drops to a low of 11.67% in the crack. The bottom line in the figure is for nickel that is at a high of 10.61% in the matrix and a low of 1.81% in the crack. Chromium, the middle plot in Figure 12.76, is highest in the crack at 37.10% and lowest in the matrix at 19.10%. It should be noted that matrix analyses for iron, chromium, and nickel very closely match the typical chemical analysis for AISI 304 SS even though the accuracy of this technique was questioned because of the surface roughness of the specimen.

Figure 12.77 shows the elemental points for silicon, sulfur, and chlorine from metal matrix to metal matrix across the crack. The top line depicts silicon content from a low of 0.83% in the metal matrix to a high of 2.22% in the crack. The middle line depicts sulfur content from a low near 0% in the matrix to a high of 1.05% in the crack. The bottom line depicts chlorine content from a low near 0% in the matrix to a high of 0.23% in the crack. The increases in sulfur level in these cracks indicate that the microbes may have caused the accumulation of sulfur in the region. The existence of significant levels of silicon and chlorine indicates the microbes may play a role in the accumulation of these elements. The level of chlorine present indicates these are chloride stress corrosion cracks, possibly near the point of initiation, based on the evidence shown in Figure 12.73 (p. 178).

Figure 12.78 shows the elemental plots for oxygen and carbon from metal matrix to metal matrix across the crack. The top line in the crack depicts a high level of 44.24% oxygen at this location and a lower level near zero level in the metal matrix. The bottom line in the crack plots carbon content from a high of 21.54% to a low of 3.84%.

The high level of oxygen in the crack shows that a large quantity of oxides is present at this location and that the microbe may be aerobic or facultatively anaerobic. The carbon level indicates that organic substances are present in the crack and on the surface of the specimen. The latter likely is, in part, the result of contamination of the surface during sanding that we tried so hard to prevent.

Further efforts were expended on the same section that was polished further using dry jeweler's rouge. No lubricant was used during these preparations with the intention of minimizing any effects from the introduction of water to the region being analyzed. An improved surface condition was achieved as is typified by the images observed and recorded.

Cracks extending deeper within the section could now be observed and were found to be filled. Figure 12.79 shows a filled crack that could not be seen in observations of previously prepared rough- sanded samples. An extension of the crack shown in Figure 12.79 is shown in Figure 12.80. This secondary phase grain-boundary deposit or crack was the focus of these efforts.

Qualitative analysis at the location (Figure 12.80) indicated the presence of silicon, sulfur and carbon in addition to the matrix elements (Figure 12.81). Qualitative step-scans were performed for carbon, oxygen, iron, sulfur, and

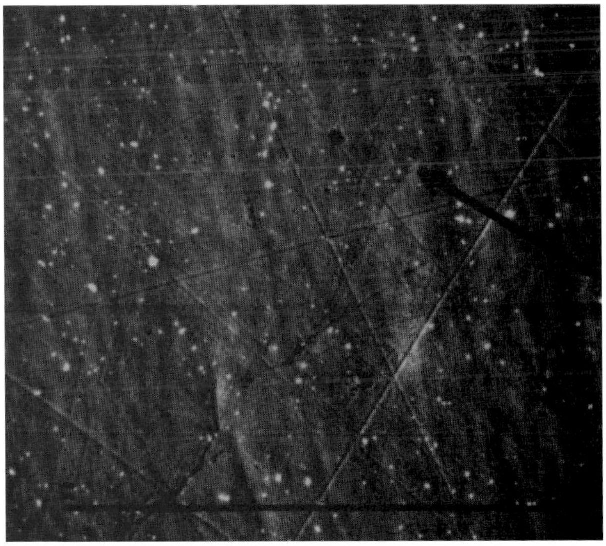

FIGURE 12.80
Photomicrograph at higher magnification of the beginning of the filled intergranular crack shown in Figure 12.79 (800X); arrow indicates region to be analyzed.

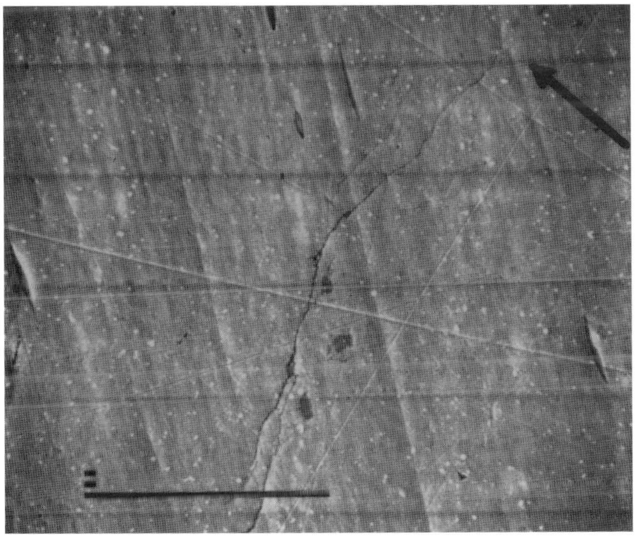

FIGURE 12.79
Photomicrograph of the beginning of a filled intergranular crack in the AISI 304 SS matrix. Arrow indicates the area of EPM investigation that was only visible following jewelers rouge polishing of the section. (400X)

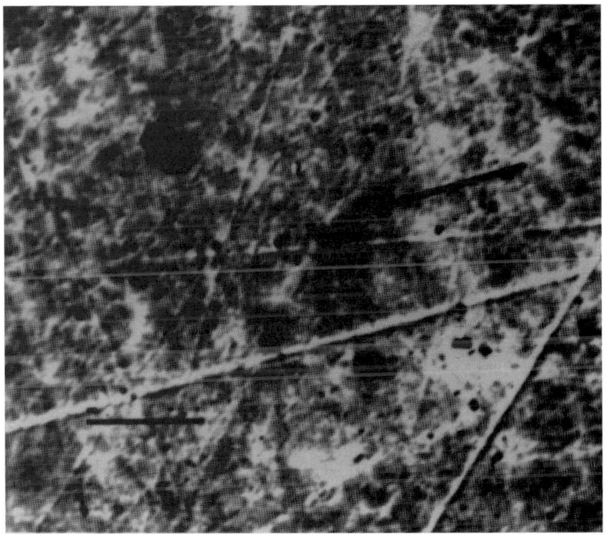

FIGURE 12.81
Electron probe microanalysis of material in intergranular crack at location shown in Figure 12.81. Scans were made at a magnification of 4000X with a step resolution of approximately 0.25 μm.

silicon. Significant increases in the levels of carbon and sulfur and a slight decrease in the iron levels were observed. The scans were made at a magnification of 4000X with a step resolution of approximately 0.25 μm.

Summary

Results indicate that intergranular corrosion is a precursor to chloride SCC in this case history of MIC of AISI 304 SS. Results show that the microbe is responsible for the formation of a deposit on the surface. It is also thought that chloride and other elements are accumulated in the corroded grain boundaries because of the action of the microbes (principally through the necessity for charge neutralization of the ionic iron present at the site of the activity of the microbes). This leads to chloride SCC of the austenitic SS.

Acknowledgments

This paper would not have been possible without the support and efforts of J.T. Schilling, E.R. Burns, and D.L. Parker. Special recognition goes to Parker for his outstanding work and dedication in electron probe microanalysis.

References

1. G. Kobrin, *Materials Performance* 15, 7(1976): p. 38.
2. R. Tatnall, J. Stoecker, R. Schultz, G. Kobrin, A. Piluso, *Materials Performance* 19, 8(1981): p. 41.
3. J. Stoecker, *Materials Performance* 23, 8(1984): p. 48.
4. D. Pope, R. Soracco, E. Wilde, CORROSION/82 paper no. 23 (Houston, TX: NACE, 1982).
5. D. Pope, D. Duquette, P. Wayner Jr., A. Johannes, *Microbiologically Influenced Corrosion: A State of the Art Review* (Columbus, Ohio: MTI, 1984).
6. J. Stoecker, *Overview of Industrial Biological Corrosion: Past, Present, and Future* (Gaithersburg, MD: NACE, 1985).
7. J. Chantereau, *Corrosion Bacterienne*, 2nd ed. (Paris, France: Techniques et Documentation, 1980).
8. J. Miller, A. Tiller, *Microbial Aspects of Metallurgy*, ed. J.D.A. Miller (Elsevier, NY: 1970).
9. W. Iverson, *Microbial Iron Metabolism*, ed. J.B. Nielands (New York, NY: Academic Press, 1974).
10. J. Postgate, *The Sulfate Reducing Bacteria* (Cambridge, UK: Cambridge University Press, 1979).
11. *Biologically Induced Corrosion*, NACE 8 (Houston, TX: NACE, 1986).

SS-6

Penetration of AISI 316 Stainless Steel Underground Pipe[13]

By J.G. Stoecker and D.H. Pope

Figure 12.82 (p. 211) shows a section of an 18-in. (45.72 cm) diameter AISI 316 SS pipe buried for underground service at a Midwest plant location. This pipeline was found to leak at several locations during attempts to place it in service after being in the ground for only seven months. Excavation of the pipeline and detailed inspection found severe pitting damage on the exterior surfaces typical of that shown in Figure 12.83 (p. 211). Microbiological analysis of the soil surrounding thc pipeline identified several species of SRB.

The solution was to replace the underground and damaged section with new AISI 316 SS pipe with an exterior coating of coal tar epoxy. Cathodic protection was applied along with the coating, plus improvements in drainage in the area and the use of a caustic-type material during backfilling. This solution has proven effective for more than two years.

[10]Orlginally published in *Process Industries Corrosion: Theory and Practice* (1986), p. 227.

SS-7

Microbiological Corrosion of Type 316 Stainless Steel

By A.A. Stein

Type 316 SS piping (0.375-in. wall) was exposed to brackish Delaware River water. Through-wall failures occurred at the weld joint. The weld filler material was type 316. Failure initiated via a small entrance pit (not shown) and developed into a bulbous interior cavity, which eventually went through-wall (Figure 12.84, p. 212).

This type of attack is typical of SRB. No corrosion occurred beneath the barnacles shown attached to the inside diameter surface of the pipe.

SS-8

MIC in Type 316 Stainless Steel Pipe with Alloy 625 Weld Metal

By A.A. Stein

Type 316 SS pipe was welded using type 308 weld filler metal. MIC occurred in the type 308 weld metal after exposure to brackish Delaware River water. These welds were then ground out and weld repaired using alloy 625 weld filler metal.

After 6 to 12 months of additional service, MIC occurred in the heat-affected zone and base metal, but avoided the alloy 625 weld metal (Figure 12.85, p. 212). Galvanic corrosion between the alloy 625 and type 316 material could not account for the corrosion observed. With additional exposures, the alloy 625 weld deposit also may become susceptible to MIC.

SS-9

MIC in Fluidized Gas Desulfurizer Plants

By A.A. Stein

SRB are usually present in wet fluidized gas desulfurizer (FGD) systems and eventually can result in corrosion of the carbon and SS materials. To date, corrosion failures due to MIC have not been widely publicized, probably due to the presence of more aggressive environments.

However, MIC has occurred in SS absorbers in nonwelded areas. High concentrations of bacteria are known to be present to such a degree that a black-brown slime can be readily observed (Figures 12.86 and 12.87, p. 212). The major effect of bacteria in FGD systems relates to plant safety concerns due to generation of H_2S and the impact of bacteria on the commercial value of gypsum.

Hydrogen sulfide is heavier than air and will collect in low points in the system. The possibility of H_2S collecting in slurry tanks can be reduced by first venting the tank and draining the slurry. Bacteria in gypsum should not be harmful, but for aesthetic reasons the gypsum can be sterilized.

Failure of 304 Stainless Steel Piping in White Water Service

By A.P. Castillo

This case history is an example of MIC from the pulp and paper industry.[1] It documents the premature failure of type 304 SS piping used in transporting paper machine white water to and from a white water storage chest.

The major source of all paper mill freshwater used to prepare the white water is a well or nearby river, and its use has always required special treatment and frequent cleanups due to its high propensity toward slime formation. Raw water treatment to combat this problem is primarily chlorination. In addition, different slimicides and biocides have been used and continue to be employed in the treated water to combat the slime problem.

Figure 12.88 (p. 213) shows the extent of corrosion damage in a two-year-old pipe section. This section of pipe was part of the seal leg completely immersed in the seal pit of one of the paper machines. Corrosion and leaks occurred for a substantial period of time before detection. Corrosive attack showed no preference for original tube welds or heat-affective zones as it varied from small craters to large holes located completely around the interior circumference. The area surrounding each hole was black.

Table 12.16 lists the chemical composition of the pipe sample. It is a typical composition for type 304. Samples of white water were taken from the clear and heavy legs of the white-water chest for analytical determination. Table 12.17 shows the concentration of all sulfur compounds and chlorides as well as an emission spectro-

TABLE 12.16
Chemical Composition of Failed Pipe Section[1]

C	0.040
Si	0.420
Mn	1.420
Cr	18.000
Ni	9.250
Mo	0.170
Cu	0.110
P	0.025
S	0.009

(Tables 12.16-12.18 reprinted with permission of the Institute of Paper Science and Technology)

TABLE 12.17
Analysis of White-Water Samples[1]

	Heavy White Water mg/L	Clear White Water mg/L
Dissolved Solids	1,400	1,380
Total Solids	1,390	1,420
Thiosulfate	22	22
Sulfate	500	510
Sulfite	2	2
Sulfide	0.4	0.3
Chloride	23.3	25.9
pH	4.25	4.40
Emission Spectrographic Analysis		
Aluminum	9.6	9.2
Calcium	73	78
Copper	0.039	0.025
Iron	1.3	1.9
Magnesium	18	23
Manganese	0.99	1.3
Silicon	8.8	11
Sodium (Major Element - Value Only Approx.)	102	83
Titanium	--	0.090

TABLE 12.18
Bacteria Tests[1]

Results of tests for SRB in two white-water samples collected and placed under test February 15, 1979:

Sample A (Stainless Valve)—24,000 SRB/mL
Sample B (Mild Steel Valve)—43 SRB/mL

Samples were tested using a three-tube MPN procedure and Starkey's medium as modified by Malek and Rizk.[2]

graphic analysis of cation concentration. The only peculiar finding in these analyses is the thiosulfate concentration (22 ppm). No other corrosive species are indicated at high enough concentrations to be of concern.

Bacteria tests were performed on samples taken near a mild steel and a stainless steel valve located close to the white water storage chest. These samples were selected because of the dark color of water first extracted from the valve. Table 12.18 shows the presence of SRB at the stainless valve location. It is important to note that sulfate reducers are present and possibly thrive in this environment.

One major reactant in the corrosion process is sulfur. Both the analyses of deposits and pit products show the presence of this element. Thiosulfate is one source of sulfur as it decomposes to bisulfite and free sulfur in acid environments or in the presence of bacteria. Of course, the metabolic activity of SRB converts sulfates to sulfides, additional sources of sulfur.

It was concluded that the primary causes of failure were:

- Low to intermittent flow of process water containing large amounts of suspended solids,
- Poor deposit control, and
- Inadequate precautionary measures to combat increased white water corrosivity due to system closure (this includes increased biological activity in various locations, e.g., clear and heavy white-water chest, temperature distribution, and white-water chemical composition profiles throughout the system).

Perhaps the most significant findings involved with the operating history of this piping failure were the nature of the water, the location of the failures, and the severity of corrosion imparted by "close-up" of the white water system. The past history of water cleanliness from the standpoint of slime and bacteria is poor, and system closure further aggravated this problem.

References

1. Institute of Paper Science and Technology (Atlanta, GA: IPST, unpublished report, 1979).
2. *Nature* 182, 4034 (1958), p. 538.

MIC Failure Analyses[(14)]

By S.W. Borenstein and P.B. Lindsay

MIC can be misdiagnosed as attack caused by conventional chloride crevice/pitting corrosion unless specialized techniques are used during the failure analysis. These techniques include *in situ* bacteria sampling of residual water, bacterial analysis of corrosion products using analytical chemistry, culture growth, and scanning electron microscopy, as well as nondestructive examination using ultrasonics and radiographic techniques. Metallographic examination can reveal MIC characteristics such as dendritic corrosion attack in weld metal. Three case histories involving AISI 304 and 304L SS are presented to illustrate the failure analysis.

Introduction

In the past few years, costly pitting failures of AISI 304 SS components have occurred in the field. Numerous articles have appeared in the corrosion literature describing failures experienced by a number of companies.[1-3] These failures have been the result of MIC and have focused attention on this problem.

MIC commonly results when residual water is left in SS systems after hydrostatic testing. Natural waters, particularly well water, contain several classes of microorganisms that can develop a biofilm in moist environments and accelerate metal corrosion. The corrosive attack in 300 series SS takes the form of pitting. Perforation of 1/8-in. (3-mm) thick SS has occurred in as little as one month.

Case 1

Pitting Corrosion of a SS Pipe Weld

This case involved SS piping installed in 1984. The piping had been hydrotested but was not yet in service when through-wall pitting was discovered. The area in which the shop-fabricated pipe spools had been stored before installation was flooded during 1980. The spool piece welds had previously been radiographed by the shop fabricator and found acceptable.

The 10-in. (250-mm) diameter line was mechanically cleaned using a cleaning pig, after which the outside diameter (OD) surface was ground to reveal the pit (Figure 12.89). After being sectioned by field personnel, the pitted area was probed with sharp instruments before

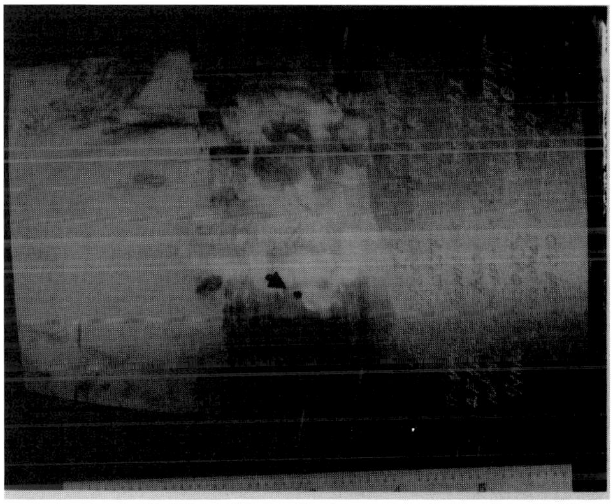

FIGURE 12.89
Photograph of the OD of an as-received section. The arrow is at the pit: Case 1. (1X)

[(14)]Originally published as CORROSION/87 paper no. 381 (Houston, TX: NACE, 1987).

being sent to the laboratory for analysis. Most corrosion deposits were removed during the field inspection.

The pipe-to-elbow weld was specified and subsequently verified to be constructed of these materials:

- Base metal (elbow): ASME SA 403, type WP 304;
- Base metal (pipe): ASME SA 312, type 304; and
- Weld metal: SFA 5.9, Class ER-308.

There was a ring of rusty discoloration around the pit and another pit on the inside diameter (ID) at the weld (Figure 12.90).

A radiographic examination (Figure 12.91) was performed on the section. The radiograph showed that the pit size was extensive in that it had a much larger subsurface volume than was apparent from the surface.

A metallographic examination revealed a small pit at the ID surface with a large cavity within the wall (Figure 12.92). The pit was ~3/8 in. (10 mm) away from the weld in the base metal of the elbow. A cross section was made through the pit in the weld, and a photomicrograph (Figure 12.93) showed selective phase attack of the weld. Figure

FIGURE 12.90
Photograph of the ID of an as-received section. The arrow is at the pit: Case 1. (1X)

FIGURE 12.91
Photograph of the radiograph of the section: Case 1. (1X)

FIGURE 12.92
Photomicrograph of a through-wall pit in the elbow, as polished: Case 1. (5X)

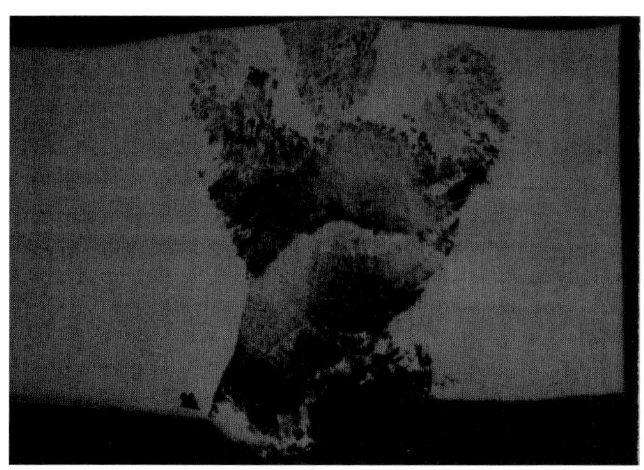

FIGURE 12.93
Photomicrograph of a pit at the weld, as polished: Case 1. (5X)

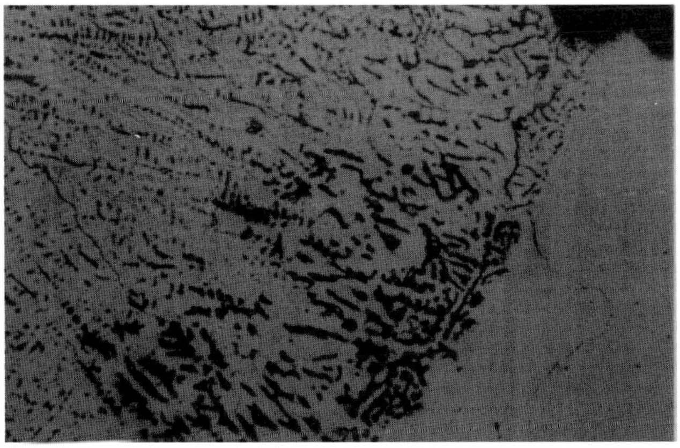

FIGURE 12.94
Photomicrograph of a pit at the weld at the location marked by the arrow in Figure 12.93. Case 1. (200X)

12.94 showed the attack in the weld metal near the cap of the weld.

The stain on the ID surface was analyzed by energy dispersive x-ray (EDX) fluorescence, which revealed particulates that were predominantly silicon with traces of sulfur, chlorine, potassium, and calcium.

A likely scenario for this failure is: It appears that untreated water resulting from the flood remained stagnant in the pipe during storage. This situation resulted in both chlorides and corrosion-causing bacteria being introduced into the system. The bacteria initiated localized corrosion cells on the inside surface. The corrosion progressed as a result of the concentration of chlorides by bacteria and the low pH generated at the base of the pits. These solutions at the base of the pits are very acidic (pH~1) and therefore drive the pit deeper and prevent the passive film of the SS from reforming inside the pit.

Case 2

Pitting Corrosion of SS Pipe

This case involved a 2-in. (50-mm) diameter SS pipe removed from a seal water supply system. Brackish service water had entered this system during construction as a result of a leaky isolation valve. In addition, the system had been primed with untreated well water. It was verified that the pipe sample met the chemical requirements of ASME SA 312, Grade 304L.

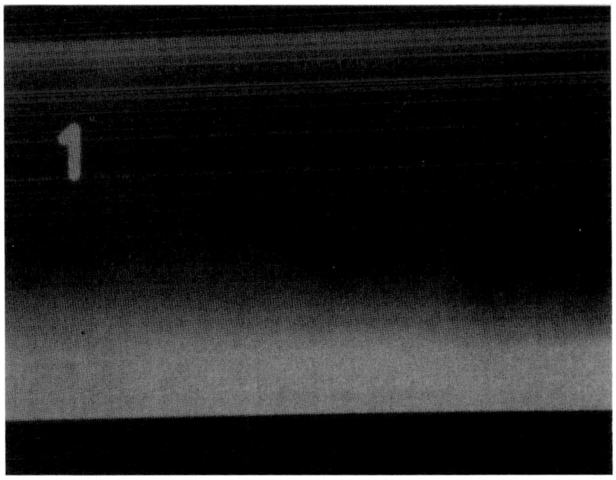

FIGURE 12.97
Radiograph of a pit: Case 2. (1X)

Visual examination of the pipe revealed a through-wall pit (Figures 12.95 and 12.96, p. 213) with minute pinhole penetrations at both the ID and OD surfaces. According to site personnel, the piping section was horizontal, and the pit was oriented in the 5 o'clock position. A rust-colored ring of corrosion deposits encircled the pit on the inside surface. The surfaces were also covered with a fine layer of sandlike residue. A radiographic examination (Figure 12.97) revealed that the subsurface pit volume was very large.

A metallographic examination of a cross section revealed what appeared to be two subsurface cavities with a series of pits branching off from each other (Figure 12.98). This type of tunneling is a frequent characteristic of pit mor-

FIGURE 12.98
Cross section of the pipe wall: Case 2. (6X)

phology in MIC failures.

Scrapings of the nodules were taken from the pipe and subjected to bacteriological analysis, which revealed *Gallionella*, an iron bacteria. Scanning electron microscopy (SEM) examination of the nodules confirmed the presence of slime, organic debris, and rod-shaped bacteria. EDX analysis of the pit and nodules indicated aluminum, silicon, potassium, sulfur, calcium, and chlorine.

A likely scenario for the situation is: A mixture of well water and brackish river water was stagnant in the pipe for some time, which resulted in both chlorides and corrosion-causing bacteria being introduced into the system. As in the previous case, the bacteria initiated localized corrosion cells on the inside surface, and the corrosion progressed as a result of the concentration of chlorides induced by bacteria and the low pH generated at the base of the pits.

Case 3

Pitting Corrosion of AISI 316L SS Pipe at Welds

A leak in a circumferential field weld was found in an 8-in. (200-mm) diameter AISI 316L SS pipe. The initial field repair to the leaking weld revealed large subsurface cavities. A visual examination of welds in the system indicated several additional leaks (Figures 12.99 and 12.100, p. 214). During the site visit, radiographs were made of five randomly picked welds (both vendor and field fabricated) in the immediate vicinity of the original pitted weld. Although no through-wall pitting had been observed in these areas, radiographs of four of the five welds showed indications in the weld. Additional radiography was done throughout the system to determine the extent of the problem. Samples were cut from two welds with a large number of indications.

A section of 14-in. (360-mm) diameter pipe submitted for evaluation was specified to be ASME SA 312, type 316L with type 308L weld metal. Chemical analysis verified that the material conformed to specification.

A visual examination revealed that the inside surface was discolored with a rusty corrosion product with a pattern of concentric rings around the pits. Surface imperfections as well as some lack of penetration to the root of the weld were detected. A radiograph (Figure 12.101) shows the extent of the defects. Direct microbiological examination of scrapings confirmed the presence of

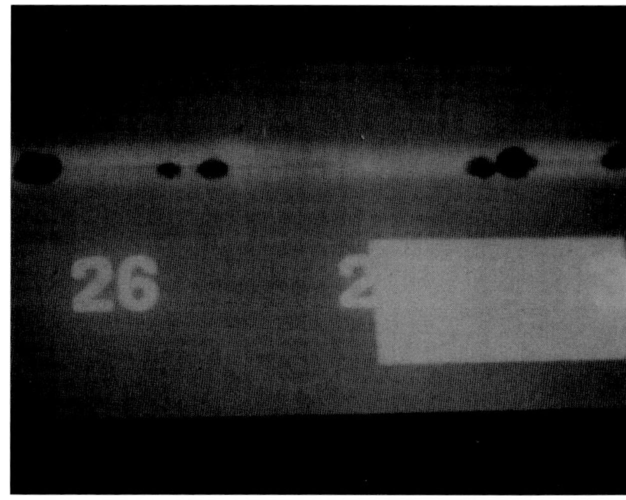

FIGURE 12.101
Radiograph of the pipe section: Case 3. (1X)

Gallionella.

A cross section (Figure 12.102) of the pipe showed ~60% through-wall penetration. A photomicrograph (Figure 12.103) showed the selective nature of the weld metal attack.

EDX was used to identify contaminants in the

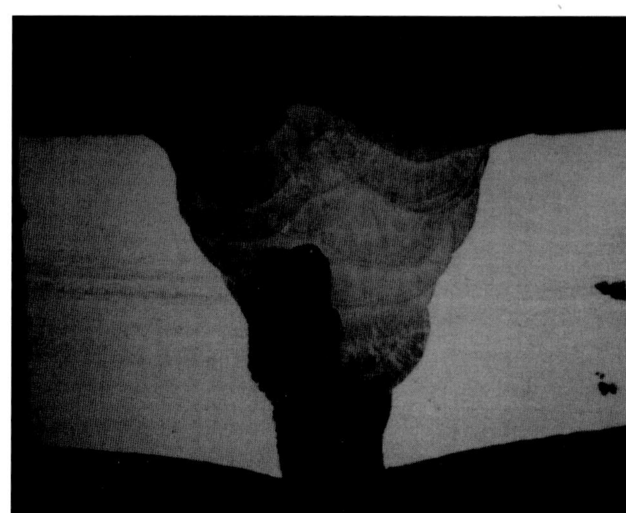

FIGURE 12.102
Cross section of the pipe weld: Case 3. (5X)

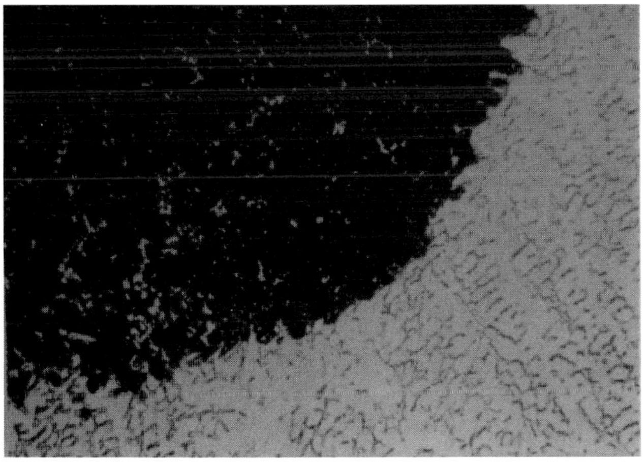

FIGURE 12.103
Selective corrosion attack in the weld: Case 3. (200X)

pits. A low peak for chloride was observed in one area. Although water quality records were unavailable, it was reported that untreated well water may have been used for flushing approximately four years before the failure occurred and that the system often sat stagnant.

Summary

Localized pitting and crevice corrosion attack in AISI 304/304L and 316/316L SS are frequently associated with aqueous chloride environments. Although such environments have undoubtedly resulted in numerous corrosion failures, the influence of microbiological activity on localized SS corrosion failures may often be obscured by the lack of bacterial analysis.

References

1. G. Kobrin, *Materials Performance* 15 7(1976): p. 38.
2. R. Tatnall, *Materials Performance* 20 9(1981): p. 32.
3. D.H. Pope, D.J. Duquette, H.A. Johannes, P.S. Wayer, CORROSION/83 paper no. 247 (Houston, TX: NACE, 1983).

MIC of Alloy 904L[(15)]

By P.J.B. Scott and M. Davies

Austenitic stainless alloy 904L (UNS N08904), comprised of approximately 20% Cr, 25% Ni, 4.5% Mo, 1.5% Cu, 0.02% max. C, and balance Fe, has been used to construct shell and tube heat exchangers for cooling sulfuric acid in brackish and seawater applications for 15 years. Experience has been excellent, with only a few isolated tube failures and reported corrosion problems.

At a recent installation in North Africa, however, tube failures occurred within only two weeks of startup accompanied by attack of welds and crevice corrosion in the waterboxes (bonnets). This phenomenon was totally unprecedented and led to a thorough investigation.

At this site, surface seawater is taken directly from an enclosed basin, passed through a coarse SS filter, and pumped underground for approximately 1.5 km (.93 mi) in concrete pipes. The water is then discharged into an open canal several hundred meters long and repumped through underground concrete pipes to the plant. The marine water enters plant equipment through rubber-lined carbon steel pipes and a further SS filter. Flow rate through the heat exchanger is in excess of 9 ft/s (3 m/s).

Chlorine is produced at the seawater intake by electrolysis of a seawater side stream, which is then reinjected as sodium hypochlorite into the main stream for continuous chlorination. Storage tanks containing hypochlorite are discharged periodically for shock treatment. Control of chlorine levels is accomplished by control of electrolytic current supplied and by matching flows of side stream to main stream. Monitoring was provided originally by occasional laboratory checks of chlorine levels at the plant.

Failure Investigation

Several potential factors were investigated. Polarization scans, chemical analyses, and metallography eliminated alloy composition and microstructure as factors and also indicated "normal" behavior of this alloy in this seawater.

Metallographic examination showed that tube pit failures were of an "ink bottle" form, i.e., with a small entry and a cavernous pit (Figure 12.104). Incipient open pits were seen over much of the tube bottom and were not associated with the tube seam weld unless this happened to be oriented at the tube bottom.

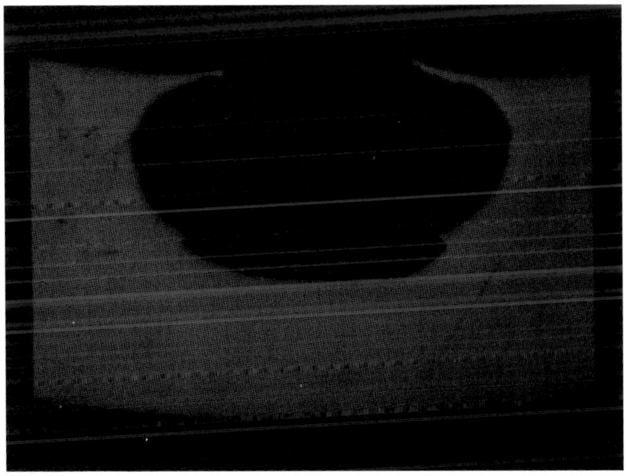

FIGURE 12.104
An "ink bottle" type pit commonly found in the alloy 904L tubes from failed heat exchanger. This type of pitting corrosion has been noted in many cases of MIC.

[(15)]Originally published in *Materials Performance* 28, 5(1989): p. 57.

FIGURE 12.105
Crevice corrosion beneath the gasket on the flange face of the water outlet nozzle of the alloy 904L heat exchanger.

FIGURE 12.106
Scanning electron micrograph of SRB from corrosion products on the waterbox of the alloy 904L cooler. Scale bar is 10 mm.

Waterboxes suffered from selective weld attack and crevice corrosion under gaskets (Figure 12.105).

Factors involved were chlorine levels, pre-service damage, and electrical interaction; water flow rate was not a contributing factor. Pre-service damage was corrected by selective retubing and repassivation treatments. Electrical effects were eliminated by plant changes.

Chlorine levels were measured by a direct colorimetric method at the heat exchanger inlet. Results indicated that high chlorine levels were present, elevating the metal potentials toward the pitting/crevicing range. The elevated potentials were determined by inserting reference electrodes through the waterboxes. Chlorine levels were reduced progressively to avoid these high potentials, and at levels approaching zero, tube failures recurred. This led to the conclusion that bacterial corrosion could also be a factor so field trials were undertaken to investigate this aspect. Details of the trials and laboratory confirmatory testing are given in the authors' CORROSION/89 paper no. 186.[1]

Standard microbiological techniques used to culture bacteria indicated the presence of large numbers of bacteria, including SRB, in the cooling water, on heat exchanger surfaces, and in areas of corrosion and corrosion products (Figure 12.106).

Intake water was observed to be very dirty, with a prominent foamed scum layer immediately adjacent to the primary screens. Laboratory analysis of this water indicated high levels of phosphates (0.7 mg/L) and nitrates (18 mg/L). Normal open ocean values for these nutrients in the tropics and subtropics approach zero.

Chlorine Analyses

Chlorine analysis was first undertaken at the site by collection of water in bottles taken to the laboratory for titration. Results of these analyses indicated variable but generally low chlorine levels. Subsequent testing with N, N-diethyl-p-phenylenediamine oxalate (DPD) colorimetric test kits at the heat exchanger indicated that previous laboratory test results erred on the low side. Sequential chlorine measurement showed that chlorine decay was variable and exponential, with a half-life ranging from 14 to 47 min.

Chlorine demand was also variable and unusually high as indicated by a large difference between total and residual free-chlorine levels. The transit period from the chlorine injection station to the coolers was 60 to 100 min. During this period, 50 to 70% of the free chlorine was lost through chlorine demand and evaporation. Demand also varied daily. For a given hypochlorite injection, the residual free-chlorine concentration was greater at night than during the day. It was anticipated, though not demonstrated, that there also would be seasonal variations in chlorine demand.

It is well known that the presence of a biofilm on SS in seawater elevates the free corrosion potential; low levels of chlorine remove this organic film from SS, thus reducing the potential.[2] However, at higher chlorine levels an alternate cathodic reaction to oxygen reduction becomes favored. At low or zero chlorine levels the probable cathodic reaction is:

$$O_2 + 2H_2O + 4e^- = 4OH\sim \qquad (1)$$

At higher chlorine levels the cathodic reaction is:

$$OCl^- + 2e^- + H_2O = Cl^- + 20H^- \qquad (2)$$

or

$$Cl_2 + 2e^- = 2\ Cl^- \qquad (3)$$

The latter reaction causes a shift to more positive potentials.

The effect of chlorine levels on metal potentials was investigated in the laboratory. Because all these samples were freshly cleaned prior to testing, the initial drop in potential due to removal of a biofilm was not seen. One part per million of free chlorine added to aerated seawater elevated the potential of alloy 904L by 270 mV at 23°C.

This positive shift in potential increases the risk of pitting and crevice crack, particularly at elevated temperatures. Therefore, it is necessary to establish chlorine doses high enough to control biological activity but not high enough to promote localized attack.

A minimum effective dose (MED) was approximated by laboratory testing. Water samples and coupons made from alloy 904L tubing were subjected to one-hour treatments of hypochlorite at various doses. Doses of 1.2 to 1.5 ppm free chlorine for one hour were found to be sufficient to kill the majority of sessile and planktonic bacteria in this system.

Examination of the extensive literature on bacterially mediated corrosion provided ample evidence that bacteria can attack SS. There appear to be no experiments, however, demonstrating bacterial corrosion of 4% and greater molybdenum-containing austenitic SS, such as alloy 904L. Therefore, to corroborate field results, laboratory experiments were conducted testing the ability of bacteria, including SRB collected at the site, to attack alloy 904L. Our results demonstrate unequivocally that alloy 904L, which contains about 4.5% molybdenum, is not immune to MIC.[1] *Bacteria were capable of both initiating and propagating pits.*

Recommended Action

Based on onsite laboratory experiments, an immediate increase in chlorine shock dose was recommended. Levels of 1.2 to 1.5 ppm free chlorine measured at the heat exchangers were implemented and were maintained for one hour daily. Maintenance levels between shock treatment were kept high enough at the intake to kill other possible corroding and fouling organisms but low enough to be near zero at the heat exchangers (0.05 to 0.10 ppm). Chlorine was kept low between shocks for two reasons:

- Bacteria and other organisms should not be able to develop chlorine-tolerant strains, and
- To permit repair of any damage to the passive film.

A monitoring program was recommended, including cultures of bulk water samples during and between shocks to detect possible variations in seasonal or other changes in bacterial biomass, and to determine general efficacy of the biocide program. In addition, heat exchanger surfaces were to be sampled during every shutdown to monitor directly sessile bacteria, especially SRB, on walls and tubes. In addition, chlorine levels were to be determined regularly both at the intake and at the heat exchangers.

Discussion

This article presents the first case history of MIC of alloy 904L. Design of the seawater system at this site did not take biological problems into account. The seawater intake collects surface water near a busy port, a probable source of the high nutrient levels measured. The open canal provided the opportunity for colonization of new bacteria after the water had been subjected to chlorination at the intake. These may enter as airborne spores, and via the soil, rain, and fresh water runoff.

Chlorine is one of the most popular biocides for control of fouling and corroding organisms. Control of chlorine at the point of injection is inappropriate because this may bear little relationship to the free-chlorine levels delivered to the plant. Chlorine decay is affected by three factors: evaporation, reactions with various chemical species in seawater, and biological chlorine demand. This latter factor is affected by the biomass of aquatic organisms, the biological oxygen demand, nutrients, temperature, and light (photosynthesis). Biological processes are, therefore, the most variable of factors affecting chlorine decay and are most important in causing unreliability of calculated chlorine doses.

Residual free chlorine must be measured independently at the plant and analyzed immediately after sampling. It should be monitored in conjunction with biological activity.

Alloy 904L is not immune to SRB attack; bacteria are capable of both initiating and propagating pits. Corrosion is relatively rapid, e.g., pits were visible during the first inspection six weeks after immersion. In addition to localized attack, bacteria appear to alter the entire surface of the steel, etching the grain boundaries and creating small pock marks on the surface. Some of the pock marks may form the site of future pit initiation. If so, creation of pits, some of which grow, occurs on the scale of one to a few micrometers, in addition to the scale more commonly observed with the naked eye or light microscope. Although biocide treatment, especially chlorine, is common in operating plants, methods of monitoring and control are often hopelessly inadequate. Lack of corrosion problems probably can be attributed more to good luck than good management. Where corrosion problems have occurred (as with other cases of biological attack), they often have been diagnosed incorrectly.

The common response to corrosion problems of this type has been to upgrade the alloy. With bacterial corrosion, however, there is no evidence that any alloy is always entirely immune. The logical upgrade from alloy 904L would be to the higher-molybdenum austenitic stainless alloys with higher critical crevice temperatures,[4] but there is no published information suggesting these are resistant to MIC attack. We are currently evaluating these and other highly alloyed grades for resistance to corrosion by bacteria.

References

1. P.J.B. Scott, M. Davies, "Microbiologically Influenced Corrosion of High Molybdenum Austenitic Stainless Steel," CORROSION/89, paper no. 186 (Houston, TX: NACE, 1989).
2. P. Gallaghner, R.E. Malpas, E.B. Shone, *Avesta Corrosion Management*, vol. 3-4, Avesta AB (Avesta, Sweden: 1986), p. 2.
3. P.D. Goodman, *Brit. Corrosion J.* 22, 1(1987): p. 56.
4. R.M. Davison, J.D. Redmond, *Materials Performance* 27, 12(1988): p. 39.

COLOR PLATES

Case Histories

FIGURE 12.1
Photograph of a 2024 aluminum alloy strip after being exposed two years in an *H. resinae* culture (pH 4.0; Eredox: 250 mV vs SCE). Fungal slime covered the part of the strip that was immersed in the water phase. (Magnification 1.5X)

FIGURE 12.3
Closeup photograph of the same *H. resinae* slime shown in Figure 12.2. (Magnification 4X)

FIGURE 12.2
Photograph of an *H. resinae* slime on a 2024 aluminum alloy strip. Light-brown deposits correspond to a mixture of corrosion products and fungal mycelia. (Magnification 1.5X)

FIGURE 12.19
Pits (white spots) with residue intact, as seen through water in cooling tower basin. Note apparent absence of (normally dark) corrosion product and discrete bacteria mounds.

FIGURE 12.20
Pits shown in Figure 12.19 after draining and drying basin. Note bright, active appearance.

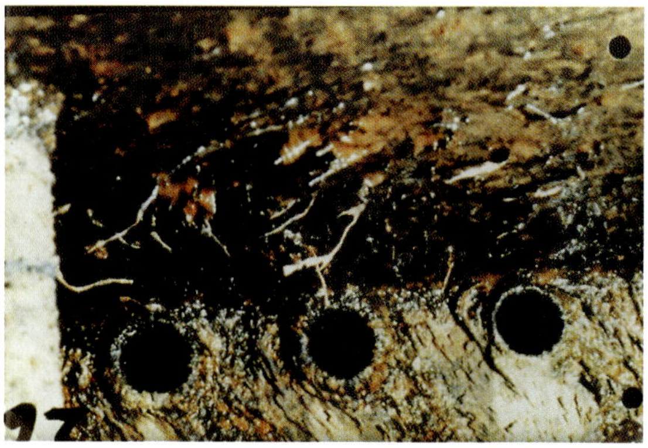

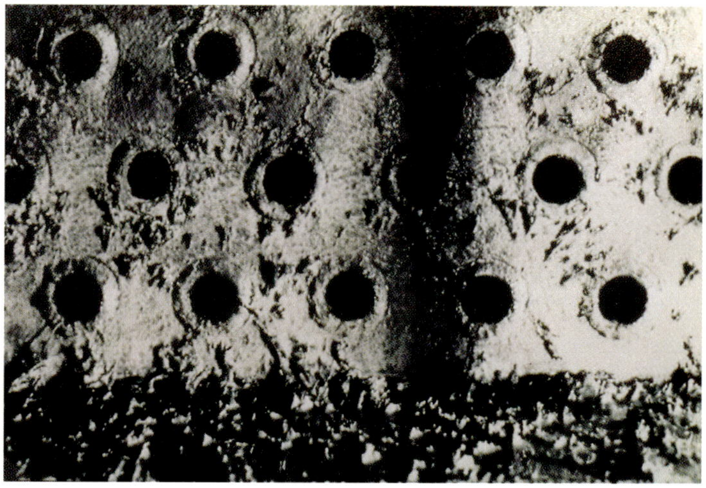

FIGURE 12.24
Reduced level of fouling. Compare to Figure 12.23. Note: Some tubercles and nodules exist on metal surfaces.

FIGURE 12.22
Upper left: Fouled heat exchanger.

FIGURE 12.23
Lower left: Fouled heat exchanger, closeup.

FIGURE 12.28
Carbon steel weir in a white water clarifier showing severe general corrosion. A heavy buildup of fiber sludge and bacterial slime had been removed prior to taking this photograph.

FIGURE 12.30
Another view of a severely corroded pump impeller. Deposits and corrosion probably occurred during periods when pump was not running.

FIGURE 12.31
Corrosion of an AISI 304 SS perforated metal screen, attributed to SRB.

FIGURE 12.32
Portion of waste treatment tank after removal of sludge. Sidewall-to-bottom junction appeared sound at this stage.

FIGURE 12.33
Closeup of sidewall-to-bottom junction in Figure 12.32 after removal of "soft" material. Such corrosion completely penetrated the channel in places.

FIGURE 12.34
Section AISI 304 SS air distribution piping showing bacteria deposits, especially on underside.

FIGURE 12.35
Underside of pipe shown in Figure 12.34 after blast cleaning.

FIGURE 12.36
AISI 304 SS piping supports and anchor bolt heads. The supports showed little attack, but many bolt heads were corroded away.

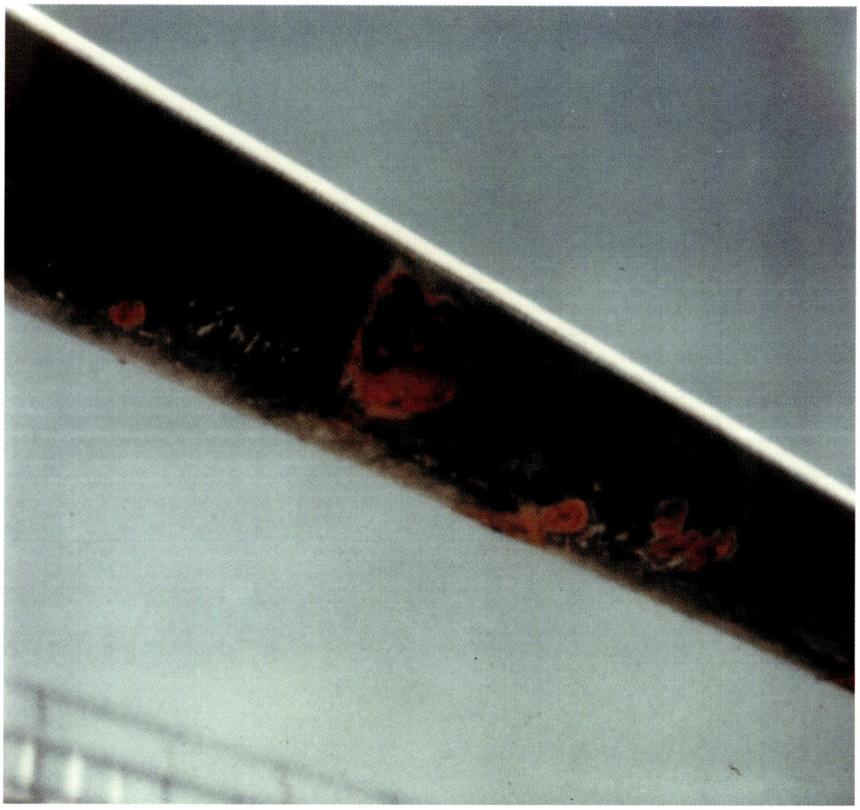

FIGURE 12.37
Internal carbon steel piping showing severe localized corrosion at voids in the coating.

FIGURE 12.38
Galvanized steel supports showing bacteria and corrosion product mounds. Severe corrosion was found under these mounds.

FIGURE 12.39
Localized corrosion of a Monel heat exchanger tube.

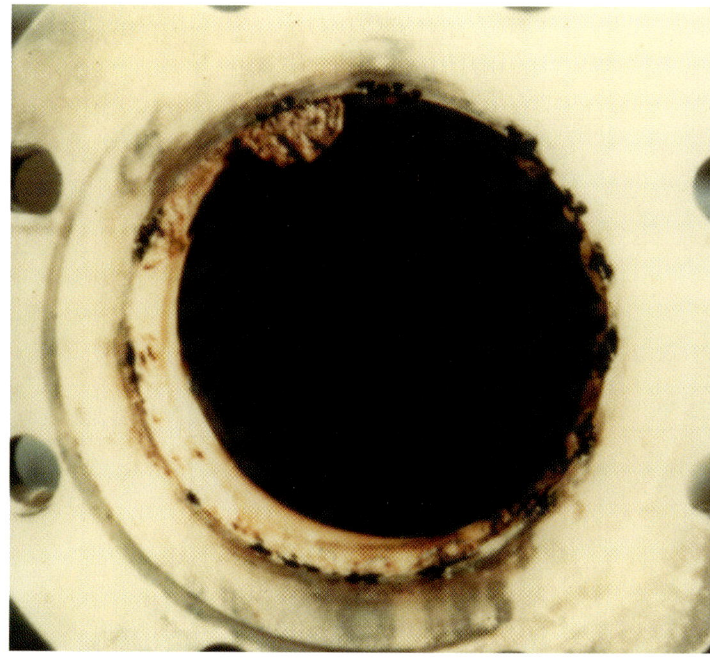

FIGURE 12.48
AISI 304 SS flange showing one remaining deposit. Numerous similar deposits were washed away when the joint was opened. This flange had been covered with an asbestos gasket and a SS blind flange.

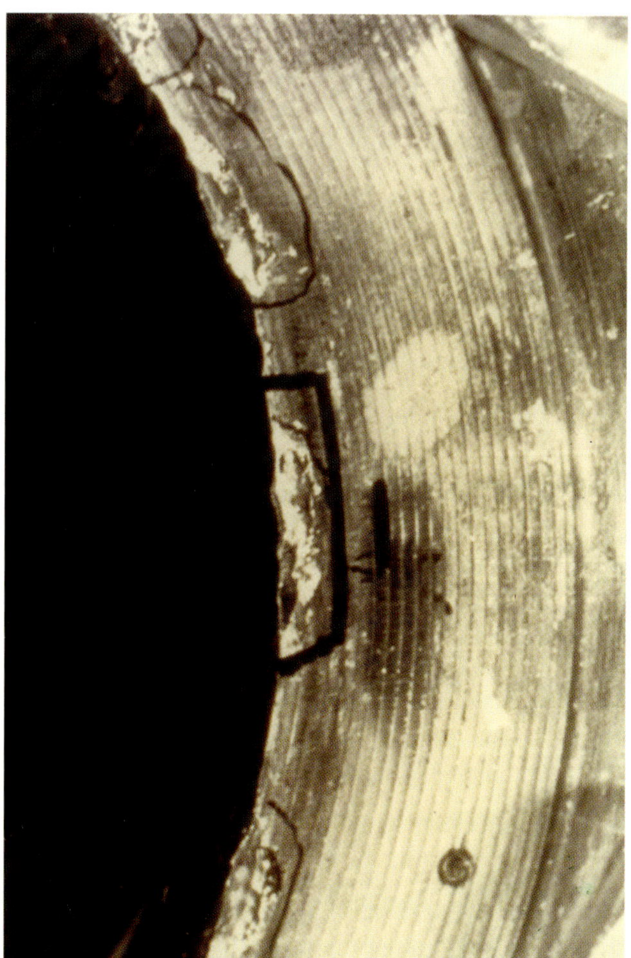

FIGURE 12.49
Closeup of flange in Figure 12.48 after cleaning. Shows open "gouging" corrosion directly under deposits.

FIGURE 12.58
Water "hangup" in cleaned tubes on condenser inlet.

FIGURE 12.50
AISI 304 SS condenser head flange showing areas of crevice corrosion, some filled with black FeS, on gasket surface. Bacteria deposits covered the inside edge of the gasket in these areas.

FIGURE 12.52
Reddish-brown deposits found along welds. (~ 3X)

FIGURE 12.55
Closeup of fish, mud, and slime in condenser inlet.

FIGURE 12.56
Slime and decayed fish pushed from tubes of condenser during cleaning.

FIGURE 12.59
Closeup of saucer-type pits. Perforation in one pit.

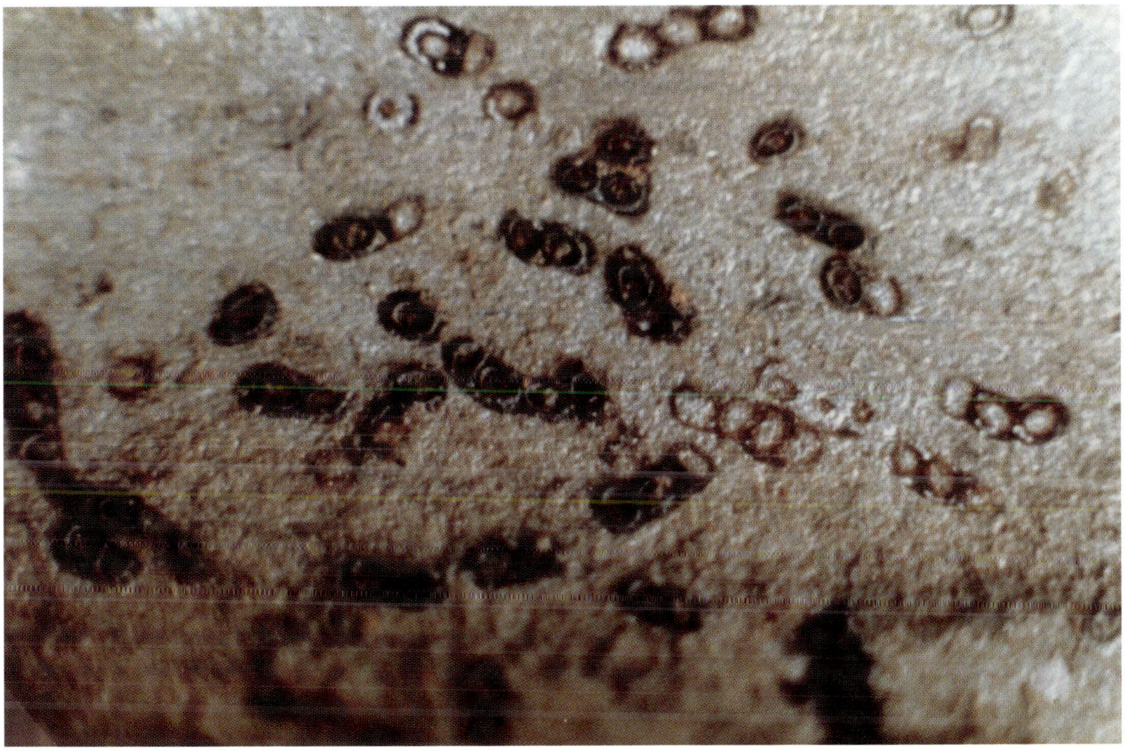

FIGURE 12.60
Tube specimen showing black spots and volcano-like formation suspected to be corrosive bacteria colonies.

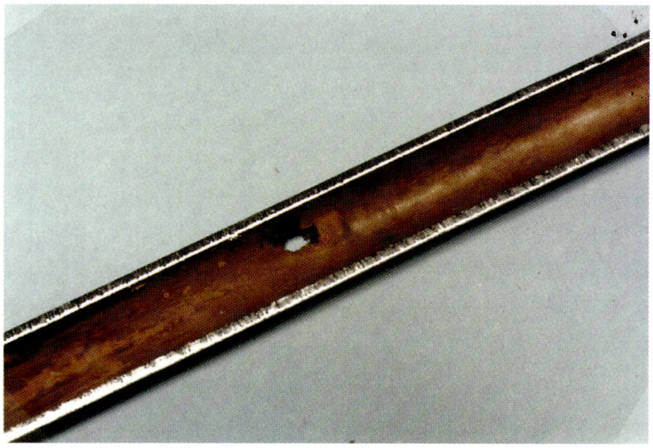

FIGURE 12.67
Corrosive penetration of ferritic-austenitic SS alloy 255 saltwater heat exchanger tube.

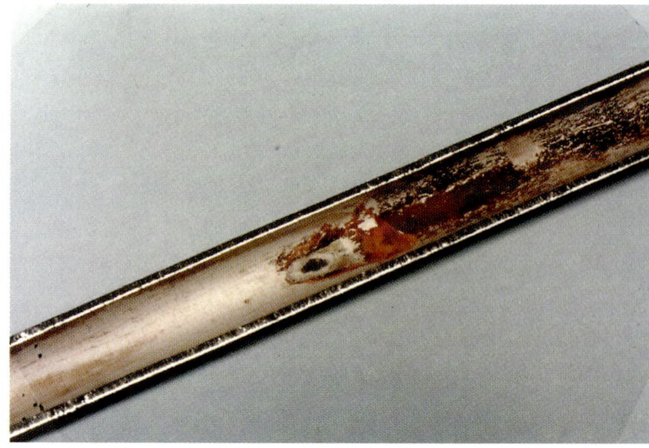

FIGURE 12.68
Localized corrosion of ferritic-austenitic SS alloy 255 saltwater heat exchanger tube.

FIGURE 12.69
View of damaged and leaking AISI 304 SS tank.

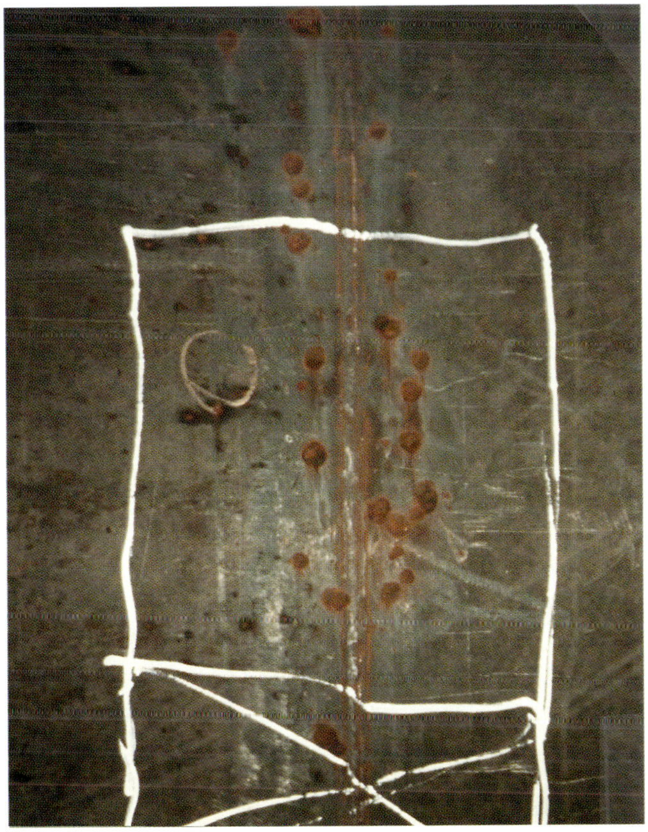

FIGURE 12.70
Typical deposits associated with the HAZ of a vertical seam weld inside the tank.

FIGURE 12.83
Above: Typical pitting found on exterior surface of underground pipeline.

FIGURE 12.82
Right: Section from 18-in. AISI 316 SS underground pipeline buried only seven months.

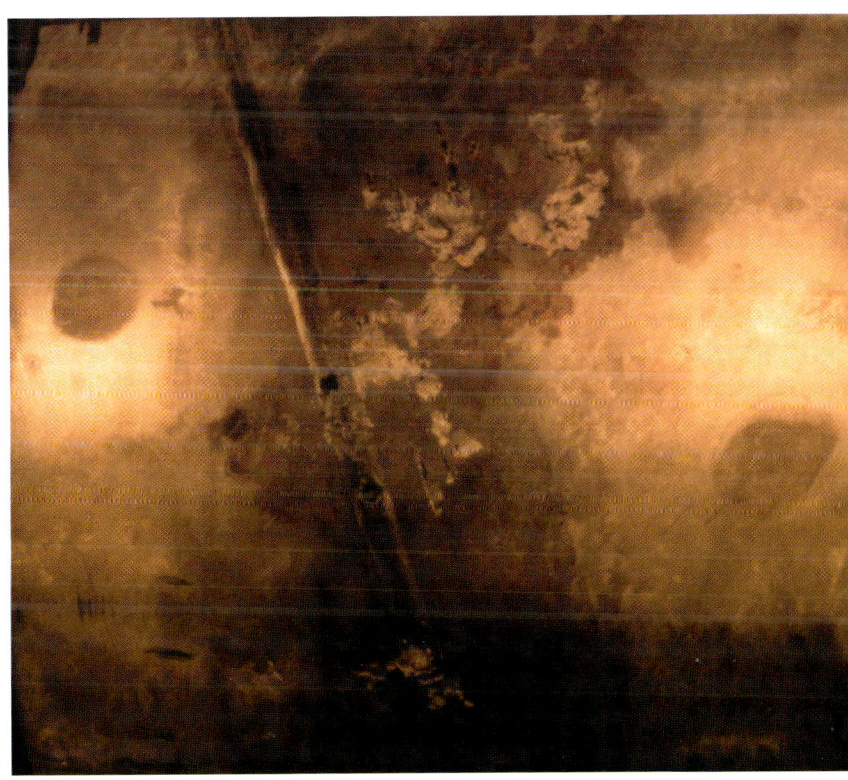

FIGURE 12.84
Type 316 SS service water piping with MIC attack after six months operation.

FIGURE 12.85
MIC in 316 SS pipe with alloy 625 weld metal. (2X)

FIGURE 12.86
MIC on test coupons in absorber tray.

FIGURE 12.87
MIC on SS absorber tray.

FIGURE 12.88
Photograph of a section of two-year old SS piping from the seal leg of a paper machine.

FIGURE 12.95
Pinhole in OD surface of pipe. Case 2.

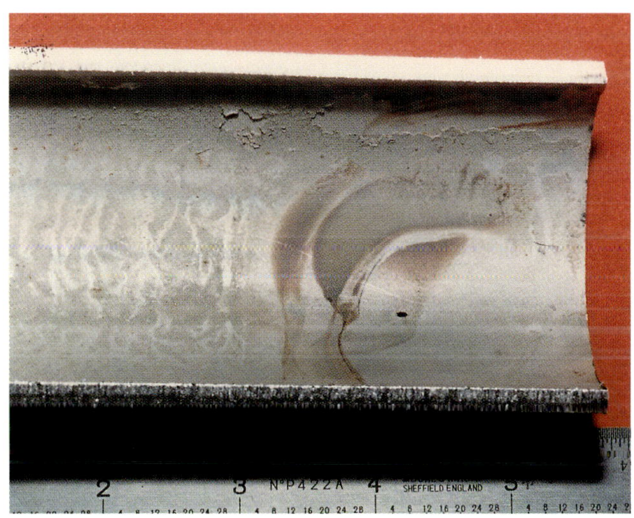

FIGURE 12.96
ID surfaces of the pipe: Case 2. (three-quarters actual size)

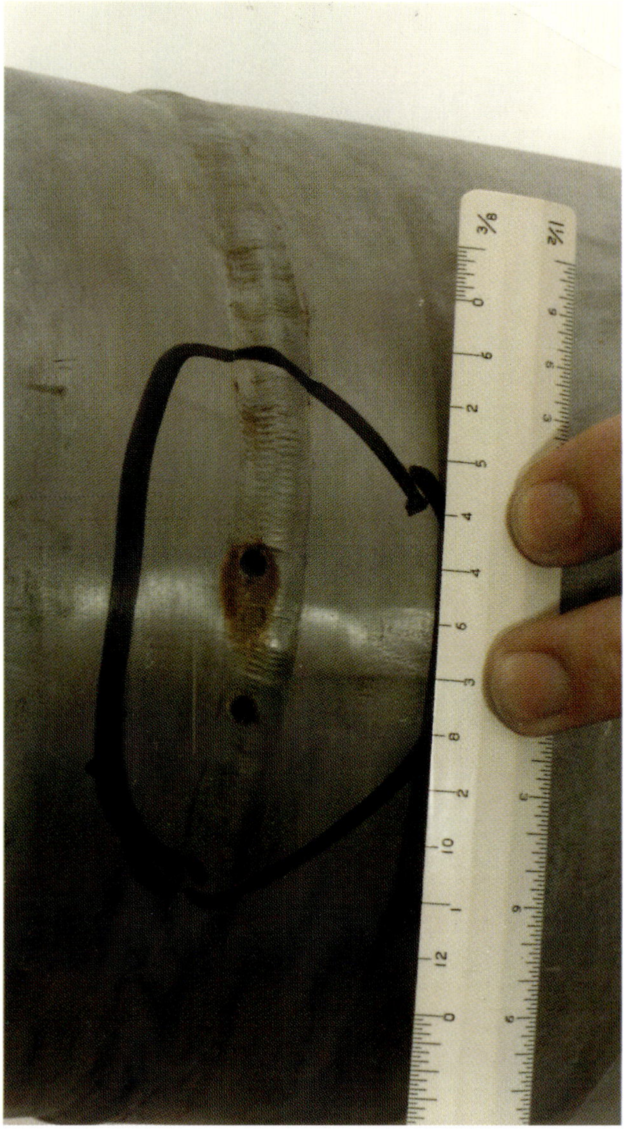

FIGURE 12.99
OD of the pipe. Through-wall pits at weld: Case 3.

FIGURE 12.100
OD of the pipe. Weeper at weld. Note water droplet at underside of pipe: Case 3.

Recommended Reading

Compiled by R.W. Lutey

1. J.R. Postgate, *The Sulfate Reducing Bacteria* (Cambridge, UK: Cambridge University Press, 1979).
2. W.P. Iverson, "Microbial Corrosion of Metals," *Advances in Applied Microbiology* 32(1987): pp. 1-37.
3. G.H. Booth, *Microbial Corrosion*. M and M Monograph CE/I (London, UK: Mills and Boon, 1971).
4. *Biologically Induced Corrosion*, Proceedings of the International Conference on Biologically Induced Corrosion, ed. S.C. Dexter (Houston, TX: NACE, 1985).
5. *The Role of Bacteria in the Corrosion of Oilfield Equipment*, TPC Pub. no. 3 (Houston, TX: NACE, 1972).
6. D.H. Pope et al., *Microbiologically Influenced Corrosion: A State of the Art Review*, MTI Pub. no. 13 (St. Louis, MO: MTI, 1989).
7. R.L. Starkey, K.M. Wright, *Anaerobic Corrosion of Iron in Soil*, American Gas Association, Vol. 37, 10 (1945): pp. 963-984.
8. G.B. Farquhar, "The Sulfate Reducing Bacteria and Oil Field Bacterial Corrosion, A Review of the Current State of the Art," NACE CORROSION/74 paper no. I (Houston, TX: NACE, 1974).
9. Institute of Petroleum, *Microbiological Up-Date for the Petroleum Industry*, ed. J.L. Shennan (London, UK: Institute of Petroleum, 1988).
10. R.E. Tatnall et al., "Testing for the Presence of Sulfate-Reducing Bacteria," *Materials Performance* 27, 8(1988): pp. 71-80.
11. Electrical Power Research Institute (EPRI), *Sourcebook of Microbiologically Influenced Corrosion in Nuclear Plants* (Prepared by G.J. Licina). EPRI NP 5580 (Palo Alto, CA: EPRI, 1988).
12. W.P. Iverson, "Corrosion of Iron and Formation of Iron Phosphide by Desulfovibrio desulfuricans," *Nature* 217 (1968): pp. 1,265-1,267.
13. E.C. Pringsheim, "Iron Bacteria," *Biological Review of the Cambridge Phil. Soc.* 24, 2(1949): pp. 200-245.
14. G. Kobrin, "Corrosion by Microbiological Organisms in Natural Waters," *Materials Performance* 15, 7(1976): pp. 38-42.
15. R.E. Tatnall, "Case Histories: Bacterial Induced Corrosion," *Materials Performance* 31, 9(1981): p. 31.
16. J.A. Costello, "The Corrosion of Metals by Microorganisms: A Literature Survey," *International Biodeterioration Bulletin* 5, 3(1969): pp. 101-118.
17. R.C. Salvarezza et al., "Mechanisms of Microbial Corrosion of Aluminum Alloys," *Corrosion* 39, 1(1983).
18. W.P. Iverson, "Mechanisms of Anaerobic Corrosion of Steel by Sulfate Reducing Bacteria," *Materials Performance* 21, 3(1984): pp. 28-30.
19. J.D.A. Miller, R.A. King, *Microbial Aspects of the Deterioration of Metals* (London, UK, New York, NY: Academic Press, 1975).
20. M.W. Mittleman, G.G. Geesey, *Biological Fouling of Industrial Water Systems: A Problem-Solving Approach* (Water Micro Associates, 1987).
21. NPL Teddington, *Microbial Corrosion* (London, UK: The Metals Society, 1983).

INDEX

G

H

I

K, L

R

S

T